METHODS AND PROTOCOLS IN FOOD SCIENCE

Series Editor
Anderson S. Sant'Ana
University of Campinas,
Campinas, Brazil

For further volumes:
http://www.springer.com/series/16556

Methods and Protocols in Food Science series is devoted to the publication of research protocols and methodologies in all fields of food science.

Volumes and chapters will be organized by field and presented in such way that the readers will be able to reproduce the experiments in a step-by-step style. Each protocol will be characterized by a brief introductory section, followed by a short aims section, in which the precise purpose of the protocol will be clarified.

Food Contaminants Toxicity

Edited by

Paula Alvito

National Institute of Health Doutor Ricardo Jorge, Department of Food and Nutrition, Lisboa, Portugal

Helena Oliveira

Centre for Environmental and Marine Studies and Department of Biology, University of Aveiro, Aveiro, Portugal

Humana Press

Editors
Paula Alvito
National Institute of Health Doutor Ricardo Jorge, Department of Food and Nutrition
Lisboa, Portugal

Helena Oliveira
Centre for Environmental and Marine Studies and Department of Biology
University of Aveiro
Aveiro, Portugal

ISSN 2662-950X ISSN 2662-9518 (electronic)
Methods and Protocols in Food Science
ISBN 978-1-0716-5095-0 ISBN 978-1-0716-5096-7 (eBook)
https://doi.org/10.1007/978-1-0716-5096-7

This Humana imprint is published by the registered company Springer Science+Business Media, LLC, part of Springer Nature.
The registered company address is: 1 New York Plaza, New York, NY 10004, U.S.A.

Preface to the Series

The Methods and Protocols in Food Science series is devoted to the publication of research protocols and methodologies in all fields of food science. The series is unique as it includes protocols developed, validated, and used by food and related scientists as well as provides theoretical basis for each protocol. Aspects related to improvements in the protocols and adaptations and further developments in the protocols may also be approached.

The Methods and Protocols in Food Science series aims to bring the most recent developments in research protocols in the field as well as very well-established methods. As such the series targets undergraduates, graduates, and researchers in the field of food science and correlated areas. The protocols documented in the series will be highly useful for scientific inquiries in the field of food sciences and presented in such a way that the readers will be able to reproduce the experiments in a step-by-step style.

Each protocol will be characterized by a brief introductory section, followed by a short aims section, in which the precise purpose of the protocol is clarified. Then, an in-depth list of materials and reagents required for employing the protocol is presented, followed by comprehensive and step-by-step procedures on how to perform that experiment. The next section brings the dos and don'ts when carrying out the protocol, followed by the main pitfalls faced and how to troubleshoot them. Finally, template results will be presented and their meaning/conclusions addressed.

The Methods and Protocols in Food Science series will fill an important gap, addressing a common complain of food scientists, regarding the difficulties in repeating experiments detailed in scientific papers. With this, the series has a potential to become a reference material in food science laboratories of research centers and universities throughout the world.

Campinas, Brazil *Anderson S. Sant'Ana*

Preface

In recent decades, global concern over food safety has grown substantially, driven by the increasing complexity of food production systems, international trade, and heightened public awareness of chemical and biological risks associated with the food supply. Among these concerns, the presence and potential toxicity of food contaminants—whether naturally occurring, introduced during processing, or resulting from environmental pollution—have taken center stage. Ensuring the safety of what we eat is no longer the sole responsibility of food producers or regulatory agencies; it is a multidisciplinary challenge that demands robust scientific frameworks, transparent methodologies, and collaborative solutions.

This book, *Methods and Protocols on Food Contaminants Toxicity*, seeks to provide a comprehensive and practical reference for scientists and graduate students engaged in the field of food safety and toxicity. It is structured around three critical pillars of food contaminant toxicity evaluation: analytical chemistry, in vitro and ex vivo toxicological approaches, and exposure assessment.

Analytical Chemistry, provides an in-depth examination of state-of-the-art techniques for detecting, identifying, and quantifying food contaminants. These chapters cover preparation, collection, and extraction protocols, for food and biological samples, and advances in instrumentation such as mass spectrometry and chromatography. Analytical reliability and sensitivity are essential not only for regulatory compliance but also for elucidating exposure pathways and supporting risk assessment.

In Vitro and Ex Vivo Toxicological Approaches explores the biological implications of food contaminants. It covers cell-based assays, organotypic models, and ex vivo testing strategies used to evaluate toxicity, mechanisms of action, and dose-response relationships. Emphasis is placed on the development of alternative methods aligned with the 3Rs principle (Replacement, Reduction, Refinement), reflecting the evolving landscape of toxicological research.

Exposure Assessment addresses the fundamental question: *how much of a contaminant is actually consumed by populations or individuals?* This includes models for dietary intake estimation, biomonitoring strategies, and probabilistic tools that help quantify exposure under various scenarios. Accurate exposure data serve as the foundation upon which toxicological relevance and regulatory thresholds are established.

Together, these three domains provide an integrated view of how food contaminants are detected, quantified, and evaluated for their potential health impacts. The methods and protocols presented in this volume have been selected and refined by experts in the field, with the goal of offering both scientific rigor and practical utility.

We hope this book serves as a valuable resource for advancing the science of food safety, improving public health outcomes, and fostering innovation in the assessment and control of food contaminants.

Lisboa, Portugal — ***Paula Alvito***
Aveiro, Portugal — ***Helena Oliveira***

Acknowledgments

The editors would like to express their sincere appreciation to all contributors who embraced the challenge of participating in this volume, sharing their expertise to enrich the discussion on food contaminants toxicity. This book was prepared within the framework of the project/grant CESAM-Centro de Estudos do Ambiente e do Mar, references UID/50017/2025 (https://doi.org/10.54499/UID/50017/2025) and LA/P/0094/2020 (https://doi.org/10.54499/LA/P/0094/2020), funded by national funds through FCT—Fundação para a Ciência e a Tecnologia I.P.

Each chapter in this volume underwent a rigorous peer-review process, and we extend our thanks to the reviewers for their careful evaluations and constructive feedback. The names of the experts who contributed to the peer review are listed below:

First Name	Last Name	Affiliation
Ana Teresa	Reis	EPIUnit—Institute of Public Health, University of Porto, Rua das Taipas 135, 4050-600, Porto, Portugal Laboratory for Integrative and Translational Research in Population Health (ITR), Rua das Taipas 135, 4050-600, Porto, Portugal Environmental Health Department, National Institute of Health, Rua Alexandre Herculano, 321, 4000-055, Porto, Portugal
Carina	Ladeira	School of Health Technology of Lisbon, Av. D. João II, Lote 4.69.01 | 1990-096 Lisboa, Portugal
Carla	Motta	Department of Food and Nutrition, National Institute of Health Doutor Ricardo Jorge, Av. Padre Cruz, 1649-016, Lisboa, Portugal
Célia	Ventura	Department of Human Genetics, Research and Development Unit, National Institute of Health Doutor Ricardo Jorge, Av. Padre Cruz, 1649-016, Lisboa, Portugal
Elsa	Vasco	Department of Food and Nutrition, National Institute of Health Doutor Ricardo Jorge, IP (INSA), Av. Padre Cruz, 1649-016 Lisboa, Portugal
Fernanda	Rosario	Centre for Environmental and Marine Studies and Department of Biology, University of Aveiro, 3810-193, Aveiro, Portugal
Flávia	Mello	MARE – Marine and Environmental Sciences Centre BIOTOX – Biomonitoring and Ecotoxicology Laboratory FCUL- Faculty of Sciences, University of Lisbon, Building C2, Floor 5, Room 2.5.50; Campo Grande, 1749-016 Lisbon, Portugal
Helena	Oliveira	Centre for Environmental and Marine Studies and Department of Biology, University of Aveiro, 3810-193, Aveiro, Portugal
Isabel	Castanheira	National Institute of Health Doutor Ricardo Jorge, IP (INSA), Av. Padre Cruz, 1649-016 Lisboa, Portugal
Letricia	Barbosa-Pereira	Department of Analytical Chemistry, Nutrition and Food Science, Faculty of Pharmacy, Rúa de Lope Gómez de Marzoa, 4, Campus Vida, 15782 Santiago de Compostela, Spain
Marc	Maresca	Aix Marseille University, CNRS, Centrale Med, ISM2, Marseille, France
Olga	Viegas	Faculty of Nutrition and Food Sciences, University of Porto, 4150-180 Porto, Portugal.
Paula	Alvito	Department of Food and Nutrition, National Institute of Health Doutor Ricardo Jorge, Av. Padre Cruz 1649-016 Lisbon, Portugal Centre for Environmental and Marine Studies, University of Aveiro, 3810-193, Aveiro, Portugal
Petru	JITARU	Laboratory for Food Safety, French Agency for Food, Environmental and Occupational Health & Safety (ANSES), 14, Pierre & Marie Curie, F-94701 Maisons-Alfort, France
Sara	Cunha	LAQV-REQUIMTE, Laboratory of Bromatology and Hydrology, Faculty of Pharmacy, University of Porto, Rua Jorge de Viterbo Ferreira n° 228, 4050-313, Porto, Portugal

Contents

Preface to the Series . . . . *v*
Preface . . . . *vii*
Acknowledgments . . . . *ix*
Contributors . . . . *xiii*

PART I ANALYTIC CHEMISTRY

1 QuEChERS Method for Pesticide Residue Analysis in Fruits and By-products 3
Ana Rita Soares Mateus, João David Teixeira, Sílvia Cruz Barros, and Ana Sanches-Silva

2 Microextraction Method for Analysis of Polybrominated Diphenyl Ethers and Their Methoxylated Congeners in Seafood 11
Mateus Henrique Petrarca, José Oliveira Fernandes, and Sara Cristina Cunha

3 Analysis of Thermic Heterocyclic Aromatic Amines in Meat Products Using One-Step SPE Followed by LC-MS/MS 25
Thayná F. Zanin, Maria V. Vicente, Fernando G. Tonin, and Alessandra V. Jager

4 ICP-MS Analysis of Inorganic Contaminants (As, Cd, Pb) in Foodstuffs Preceded by Closed Vessel Microwave Acid Digestion 35
Marta Ventura, Andreia Rego, Inês Delgado, Sandra Gueifão, and Inês Coelho

5 Ultra-Performance Liquid Chromatography-Tandem Mass Spectrometry (UPLC-MS/MS) for the Determination of Acrylamide in Grains and Grain-Based Products 47
Inês Delgado, Susana Jesus, Sandra Copeto, and Inês Coelho

6 Determination of Nitrate and Nitrite in Beet Chips Using UV-Vis Spectrophotometry 57
David Silva da Costa, Ana Beatriz Porto Fernandes da Silva, Gabriel de Souza Macedo, Luiza Saraiva Bebiano, Camila Akemi Akiyama, Patricia Aparecida de Campos Braga, Juliana Azevedo Lima Pallone, and Adriana Pavesi Arisseto Bragotto

7 Heavy Metal Phytotoxicity Evaluation: Method for Antioxidant Profiling in Leaves 71
Marcelo D. Catarino, Susana M. Cardoso, Sónia Silva, and Maria Celeste Dias

8 ELISA-Based Protocols for Mycotoxin Detection Across Different Matrices in Africa 81
Wilfred Angie Abia and Herman Philipe Nfombouot Njitoyap

Part II In Vitro & Ex Vivo Toxicological Approaches

9 From Protocol to Analysis: Performing and Interpreting the MTT Assay ... 103
Andreia Rego, José Furtado, José A. Silva, Paula Alvito, and Carla Motta

10 Advanced In Vitro Models to Assess Contaminants Toxicity by Ingestion... 113
Fernanda Rosário and Helena Oliveira

11 Integrated In Vitro Genotoxicity Testing Protocols for Safety Assessment of Food Nanomaterials After Simulated Human Digestion... 121
Henriqueta Louro, Nádia Vital, Maria Cardoso, Paula Alvito, and Maria João Silva

12 Adsorption of Potentially Toxic Elements onto Nanomaterials in Aqueous and Biological Media: Supporting Chemical Co-exposure Toxicity Assessment ... 141
Ana Teresa Reis, João Paulo Teixeira, and Cláudia Batista Lopes

13 Human Ingestion of Micro-/Nanoplastics: A Simplified In Vitro Digestion Protocol for Cytotoxicity and Risk Assessment ... 157
Verónica Bastos and Helena Oliveira

14 Porcine Explants and Slices: Original Models to Analyze the Effects of Mycotoxins ... 167
Philippe Pinton, Delphine Payros, Laura Soler-Vasco, and Isabelle P. Oswald

Part III Exposure Assessment

15 Total Diet Study: An Example for Application of a Harmonized Public Health Methodology for Assessing Dietary Exposure to Chemical Contaminants... 181
Elsa Vasco, Luísa Oliveira, and M. Graça Dias

16 Food Contaminants Exposure Assessment Using Random Forest Model 195
Sofia Almeida Costa, Milton Severo, Catarina Carvalho, Carla Lopes, and Duarte Torres

Index ... *205*

Contributors

WILFRED ANGIE ABIA • *Department of Biochemistry, Faculty of Science, University of Yaounde 1, Yaounde, Cameroon; Agri-Food Safety and One Health Agency (AFS1HA), Yaounde, Cameroon*

CAMILA AKEMI AKIYAMA • *Faculty of Food Engineering, Universidade Estadual de Campinas (UNICAMP), Campinas, Brazil*

PAULA ALVITO • *Department of Food and Nutrition, National Institute of Health Doutor Ricardo Jorge, Lisbon, Portugal; Centre for Environmental and Marine Studies, University of Aveiro, Aveiro, Portugal*

SÍLVIA CRUZ BARROS • *INIAV—National Institute for Agricultural and Veterinary Research, I.P., Rua dos Lagidos, Lugar da Madalena, Vila do Conde, Portugal*

VERÓNICA BASTOS • *Centre for Environmental and Marine Studies, Department of Biology, University of Aveiro, Aveiro, Portugal*

LUIZA SARAIVA BEBIANO • *Faculty of Food Engineering, Universidade Estadual de Campinas (UNICAMP), Campinas, Brazil*

ADRIANA PAVESI ARISSETO BRAGOTTO • *Faculty of Food Engineering, Universidade Estadual de Campinas (UNICAMP), Campinas, Brazil*

MARIA CARDOSO • *Department of Human Genetics, National Institute of Health Doutor Ricardo Jorge, Lisbon, Portugal*

SUSANA M. CARDOSO • *LAQV-REQUIMTE, Department of Chemistry, University of Aveiro, Campus Universitário de Santiago, Aveiro, Portugal*

CATARINA CARVALHO • *EPIUnit ITR – Institute of Public Health, University of Porto, Porto, Portugal; Faculty of Nutrition and Food Sciences, University of Porto, Porto, Portugal*

MARCELO D. CATARINO • *LAQV-REQUIMTE, Department of Chemistry, University of Aveiro, Campus Universitário de Santiago, Aveiro, Portugal*

INÊS COELHO • *Department of Food and Nutrition, National Institute of Health Doutor Ricardo Jorge (INSA IP), Lisbon, Portugal*

SANDRA COPETO • *Department of Food and Nutrition, National Institute of Health Doutor Ricardo Jorge (INSA IP), Lisbon, Portugal; Applied Molecular Biosciences Unit (UCIBIO), Department of Chemistry, NOVA School of Science and Technology, Nova University Lisbon, Quinta da Torre, Caparica, Portugal*

SOFIA ALMEIDA COSTA • *EPIUnit ITR – Institute of Public Health, University of Porto, Porto, Portugal*

SARA CRISTINA CUNHA • *LAQV-REQUIMTE, Laboratory of Bromatology and Hydrology, Faculty of Pharmacy, University of Porto, Porto, Portugal*

DAVID SILVA DA COSTA • *Faculty of Food Engineering, Universidade Estadual de Campinas (UNICAMP), Campinas, Brazil*

ANA BEATRIZ PORTO FERNANDES DA SILVA • *Faculty of Food Engineering, Universidade Estadual de Campinas (UNICAMP), Campinas, Brazil*

PATRICIA APARECIDA DE CAMPOS BRAGA • *Faculty of Food Engineering, Universidade Estadual de Campinas (UNICAMP), Campinas, Brazil*

GABRIEL DE SOUZA MACEDO • *Faculty of Food Engineering, Universidade Estadual de Campinas (UNICAMP), Campinas, Brazil*

INÊS DELGADO • *Department of Food and Nutrition, National Institute of Health Doutor Ricardo Jorge (INSA IP), Lisbon, Portugal*
M. GRAÇA DIAS • *Department of Food and Nutrition, National Institute of Health Doutor Ricardo Jorge, IP (INSA), Lisbon, Portugal*
MARIA CELESTE DIAS • *Department of Life Sciences, Centre for Functional Ecology (CFE)–Science for People & the Planet, Associate Laboratory TERRA, University of Coimbra, Coimbra, Portugal*
JOSÉ OLIVEIRA FERNANDES • *LAQV-REQUIMTE, Laboratory of Bromatology and Hydrology, Faculty of Pharmacy, University of Porto, Porto, Portugal*
JOSÉ FURTADO • *Department of Human Genetics, National Institute of Health Doutor Ricardo Jorge, Lisbon, Portugal*
SANDRA GUEIFÃO • *Department of Food and Nutrition, National Institute of Health Doutor Ricardo Jorge (INSA IP), Lisbon, Portugal*
ALESSANDRA V. JAGER • *School of Pharmaceutical Sciences at Ribeirão Preto, University of São Paulo, Ribeirão Preto, SP, Brazil*
SUSANA JESUS • *Department of Food and Nutrition, National Institute of Health Doutor Ricardo Jorge (INSA IP), Lisbon, Portugal; MARE– Marine and Environmental Sciences Centre, ARNET– Aquatic Research Network Associate Laboratory, NOVA School of Science and Technology, NOVA University Lisbon, Caparica, Portugal*
CARLA LOPES • *EPIUnit ITR – Institute of Public Health, University of Porto, Porto, Portugal; Department of Public and Forensic Health Sciences, and Medical Education, Faculty of Medicine, University of Porto, Porto, Portugal*
CLÁUDIA BATISTA LOPES • *Department of Chemistry and Aveiro Institute of Materials (CICECO), University of Aveiro, Aveiro, Portugal*
HENRIQUETA LOURO • *Department of Human Genetics, National Institute of Health Doutor Ricardo Jorge, Lisbon, Portugal; UID/04923 – Comprehensive Health Research Centre, NOVA Medical School, Universidade NOVA de Lisboa, Lisbon, Portugal*
ANA RITA SOARES MATEUS • *INIAV—National Institute for Agricultural and Veterinary Research, I.P., Rua dos Lagidos, Lugar da Madalena, Vila do Conde, Portugal; LAQV, REQUIMTE, Laboratory of Bromatology and Pharmacognosy, Faculty of Pharmacy, University of Coimbra, Coimbra, Portugal*
CARLA MOTTA • *Department of Food and Nutrition, National Institute of Health Doutor Ricardo Jorge, Lisbon, Portugal*
HERMAN PHILIPE NFOMBOUOT NJITOYAP • *Department of Biochemistry, Faculty of Science, University of Yaounde 1, Yaounde, Cameroon*
HELENA OLIVEIRA • *Centre for Environmental and Marine Studies, Department of Biology, University of Aveiro, Aveiro, Portugal*
LUÍSA OLIVEIRA • *Department of Food and Nutrition, National Institute of Health Doutor Ricardo Jorge, IP (INSA), Lisbon, Portugal*
ISABELLE P. OSWALD • *Toxalim (Research Centre in Food Toxicology), Université de Toulouse, INRAE, ENVT, INP-Purpan, UPS, Toulouse, France*
JULIANA AZEVEDO LIMA PALLONE • *Faculty of Food Engineering, Universidade Estadual de Campinas (UNICAMP), Campinas, Brazil*
DELPHINE PAYROS • *Toxalim (Research Centre in Food Toxicology), Université de Toulouse, INRAE, ENVT, INP-Purpan, UPS, Toulouse, France; National Veterinary School of Toulouse (ENVT), INRAE, UMR 1225, IHAP, Université de Toulouse, Toulouse, France*

MATEUS HENRIQUE PETRARCA • *LAQV-REQUIMTE, Laboratory of Bromatology and Hydrology, Faculty of Pharmacy, University of Porto, Porto, Portugal*

PHILIPPE PINTON • *Toxalim (Research Centre in Food Toxicology), Université de Toulouse, INRAE, ENVT, INP-Purpan, UPS, Toulouse, France*

ANDREIA REGO • *Department of Food and Nutrition, National Institute of Health Doutor Ricardo Jorge, Lisbon, Portugal; Department of Chemical Engineering, Instituto Superior Técnico, Lisbon, Portugal*

ANA TERESA REIS • *EPIUnit—Institute of Public Health, University of Porto, Porto, Portugal; Laboratory for Integrative and Translational Research in Population Health (ITR), Porto, Portugal; Environmental Health Department, National Institute of Health, Porto, Portugal*

FERNANDA ROSÁRIO • *Centre for Environmental and Marine Studies, Department of Biology, University of Aveiro, Aveiro, Portugal*

ANA SANCHES-SILVA • *CECA—Centre for Animal Science Studies, ICETA, Porto, Portugal; University of Coimbra, Faculty of Pharmacy, Coimbra, Portugal; Al4AnimalS—Associate Laboratory for Animal and Veterinary Sciences, Lisbon, Portugal*

MILTON SEVERO • *EPIUnit ITR – Institute of Public Health, University of Porto, Porto, Portugal; Abel Salazar Institute of Biomedical Sciences, University of Porto, Porto, Portugal*

JOSÉ A. SILVA • *Centre for Structural Chemistry, Institute of Molecular Sciences, Department of Chemical Engineering, Instituto Superior Técnico, University of Lisbon, Lisbon, Portugal*

MARIA JOÃO SILVA • *Department of Human Genetics, National Institute of Health Doutor Ricardo Jorge, Lisbon, Portugal; UID/04923 – Comprehensive Health Research Centre, NOVA Medical School, Universidade NOVA de Lisboa, Lisbon, Portugal*

SÓNIA SILVA • *LAQV-REQUIMTE, Department of Chemistry, University of Aveiro, Campus Universitário de Santiago, Aveiro, Portugal*

LAURA SOLER-VASCO • *Toxalim (Research Centre in Food Toxicology), Université de Toulouse, INRAE, ENVT, INP-Purpan, UPS, Toulouse, France*

JOÃO DAVID TEIXEIRA • *INIAV—National Institute for Agricultural and Veterinary Research, I.P., Rua dos Lagidos, Lugar da Madalena, Vila do Conde, Portugal; CECA—Centre for Animal Science Studies, ICETA, Porto, Portugal; Department of Chemistry, Portugal and University of Minho, Braga, Portugal*

JOÃO PAULO TEIXEIRA • *EPIUnit—Institute of Public Health, University of Porto, Porto, Portugal; Laboratory for Integrative and Translational Research in Population Health (ITR), Porto, Portugal; Environmental Health Department, National Institute of Health, Porto, Portugal*

FERNANDO G. TONIN • *School of Animal Science and Food Engineering, University of São Paulo, Pirassununga, SP, Brazil*

DUARTE TORRES • *EPIUnit ITR – Institute of Public Health, University of Porto, Porto, Portugal; Faculty of Nutrition and Food Sciences, University of Porto, Porto, Portugal*

ELSA VASCO • *Department of Food and Nutrition, National Institute of Health Doutor Ricardo Jorge, IP (INSA), Lisbon, Portugal*

MARTA VENTURA • *Department of Food and Nutrition, National Institute of Health Doutor Ricardo Jorge (INSA IP), Lisbon, Portugal; MARE – Marine and Environmental Sciences Centre, ARNET – Aquatic Research Network Associate Laboratory, NOVA School of Science and Technology, NOVA University Lisbon, Caparica, Portugal*

MARIA V. VICENTE • *School of Pharmaceutical Sciences at Ribeirão Preto, University of São Paulo, Ribeirão Preto, SP, Brazil*
NÁDIA VITAL • *Department of Human Genetics, National Institute of Health Doutor Ricardo Jorge, Lisbon, Portugal; UID/04923 – Comprehensive Health Research Centre, NOVA Medical School, Universidade NOVA de Lisboa, Lisbon, Portugal*
THAYNÁ F. ZANIN • *School of Pharmaceutical Sciences at Ribeirão Preto, University of São Paulo, Ribeirão Preto, SP, Brazil*

Part I

Analytic Chemistry

Chapter 1

QuEChERS Method for Pesticide Residue Analysis in Fruits and By-products

Ana Rita Soares Mateus, João David Teixeira, Sílvia Cruz Barros, and Ana Sanches-Silva

Abstract

The QuEChERS (Quick, Easy, Cheap, Effective, Rugged, and Safe) method is a widely used approach for the determination of pesticide residues in food matrices, particularly in fruits. Developed by Anastassiades and coworkers, this method has revolutionized sample preparation by simplifying extraction and cleanup steps while ensuring high analyte recovery and minimal matrix effects. It relies on liquid-liquid partitioning with acetonitrile, followed by dispersive solid-phase extraction (d-SPE) to remove impurities, making it an efficient and cost-effective alternative to traditional methods.

This chapter provides a detailed description of the QuEChERS procedure, including key variations, optimization strategies, and analytical considerations for pesticide residue detection in fruit-based samples. It also discusses its advantages, such as reduced solvent consumption and adaptability to different analytical techniques, including gas and liquid chromatography coupled with mass spectrometry. The method's role in ensuring food safety and compliance with regulatory limits is highlighted, along with recent advancements and future perspectives for its application in the agro-food industry.

Key words QuEChERS, Residues of pesticides, Fruits

1 Introduction

The consumption of fresh and processed fruits has been steadily increasing in this millennium, with fruit consumption per capita per year reaching 86.4 kg in 2021 [1]. Therefore, it is important to ensure that there are sufficient fruits being produced, and in order to guarantee that, pesticides have been used in agriculture to protect crops from various diseases and improve crops yield [2]. However, food safety measures must be followed, and the improper or persistent use of pesticides can lead to their presence on the fruits [3], and because it is well known that pesticide residues have potential health effects like neurotoxicity [4, 5] and carcinogenicity

Paula Alvito and Helena Oliveira (eds.), *Food Contaminants Toxicity*, Methods and Protocols in Food Science,
https://doi.org/10.1007/978-1-0716-5096-7_1,

[6], regulatory agencies demand rigorous monitoring of pesticide levels in food products.

The extraction techniques mainly used for pesticides residues are liquid-liquid extraction (LLE), solid-phase extraction (SPE) and matrix solid-phase dispersion (MSPD), but while effective, these methods are time consuming, require a great amount of solvents, and are very laborious. In response to these limitations, Anastassiades et al. [7] developed the QuEChERS (acronym of Quick, Easy, Cheap, Effective, Rugged, and Safe) method to be applied in food matrices that have high amounts of water (>75%) and low-fat content. However, it was later proven that the method could be adapted for low-moisture and high-fat matrices by modifying the hydration step (e.g., adding water prior to extraction), adjusting salt composition, or incorporating additional cleanup sorbents (such as C18, Z-Sep, or EMR-Lipid) to effectively remove lipids and nonpolar interferences [8].

The method essentially involves a two-step process: (1) an extraction with a mixture of salts, separating the pesticides from the unwanted polar compounds to improve recoveries of pH-dependent analytes and (2) a dispersive solid-phase extraction (d-SPE) to eliminate the remaining interfering compounds like lipids and sugars.

In line with the above, the QuEChERS methods present a wide range of advantages when compared to the traditional methods, like the use of smaller amounts of solvents, high recovery, simplicity, and efficiency, as well as a rapid execution time. Another advantage of this method is that it provides good recoveries for different classes of pesticide residues. The use of the solvent acetonitrile in the final extract allows flexibility in the choice of techniques applied: gas or liquid chromatography. In this context, multiple detection methods have become easier and faster, leading to good results for a wide range of compounds.

This chapter aims to provide a comprehensive overview of the QuEChERS method applied for the analysis of pesticide residues in fruits and by-products. This chapter also contains a thorough description of the preparation of every reagent utilized as well as the application of the method, with several notes that cover variations and optimizations for different food matrices.

2 Materials

1. Prepare all solutions using analytical grade reagents: methanol, acetonitrile (both HPLC gradient grade), toluene, acetone, ethanol, ethyl acetate, n-hexane, formic acid, and ultrapure water (obtained by purifying water in Milli-Q plus system with a resistivity of 18.2 MΩ × cm at 25 °C). Prepare and store all solutions in amber vials in the dark at −20 °C, for up

to at least 3 years, to reduce the rate of decomposition. Before use, allow the solutions to equilibrate at room temperature for 15 minutes. Diligently follow all waste disposal regulations when disposing waste materials.

2.1 Mixture of Salts for the Liquid-Liquid Partitioning Step

1. Per sample, the mixture of salts for the liquid-liquid partitioning step contains 4 g $MgSO_4$, 1 g NaCl, 1 g of trisodium citrate dihydrate, and 0.5 g of disodium hydrogen citrate sesquihydrate. Thus, for 20 samples, weigh 80 g of magnesium sulfate ($MgSO_4$), 20 g of sodium chloride (NaCl), 20 g trisodium citrate dihydrate, and 10 g of disodium hydrogen citrate sesquihydrate.
2. Carefully mix all the four reagents to ensure a complete homogenization.

2.2 Mixture of Salts for Cleanup Step

1. Per sample, the mixture of salts for clean-up step contains 900 mg of anhydrous $MgSO_4$ and 150 mg of PSA. Thus, for 20 samples, weigh 18 g of anhydrous magnesium sulfate ($MgSO_4$) and 3 g of primary secondary amine (PSA).
2. Carefully mix all the four reagents to ensure a complete homogenization.

2.3 Pesticides Standards

1. Stock Solutions of Pesticides: Each pesticide standard is dissolved in toluene, acetone, ethanol, ethyl acetate, methanol, n-hexane, or acetonitrile, depending on the solubility of the compound, at a concentration of 1 μg/mL (*see* **Note 1**).
2. Mix Solution of Pesticides: The stock solutions are subsequently used to prepare mixture of all pesticide's standards at 5 ng/μL in acetonitrile.
3. Working Solutions: The Mix Solution is subsequently used to prepare different working solutions for calibrations, prepared in acetonitrile, at 2.5 ng/μL and 1 ng/μL.

2.4 Internal Standards (ISs)

1. Stock Solutions of IS: Internal standards included triphenylphosphate-(TPP) and dinitrocarbanilide or 1,3-bis (4-nitrophenyl)urea (DNC). Individual stock solutions of TPP and DNC are prepared at a concentration of 5 μg/mL, dissolved in acetonitrile.
2. Mix Solution: The stock solutions of TPP and DNC are subsequently used to prepare mixture of internal standards at 2.5 ng/μL in acetonitrile.
3. Working Solution: The Mix Solution is subsequently used to prepare different working solutions, at 1 ng/μL, prepared in acetonitrile.

3 Methods

1. Weight 10 g of sample in a 50 mL polypropylene Falcon tube (*see* **Note 2**).
2. Add 10 mL of acetonitrile to the Falcon tube with the sample (*see* **Note 3**).
3. Vortex the sample and the extractant solvent for 1 min.
4. Add 6.5 g of mixture of extraction salts for the liquid-liquid partitioning step (*see* Subheading 2.1, *see* **Notes 4** and **5**).
5. Shake vigorously by hand, immediately after adding the salts, and then, vortex for 1 min (*see* **Note 6**).
6. Centrifuge the samples at 12,669× *g* for 10 min at 4 °C.
7. In a 25 mL polypropylene Falcon tube, weigh 1.05 g of cleanup mixture (see Subheading 2.2, *see* **Notes 5** and 7).
8. Add 6 mL of the supernatant (after **step 6**) to the 15 mL Falcon tube containing the cleanup mixture.
9. Vigorously shake by hand the tubes for 1 min.
10. Centrifuge the samples at 12,669× *g* for 2 min at 4 °C.
11. Transfer 1 mL of the cleaned extract to an Eppendorf.
12. Add 220 μL of acetonitrile to the Eppendorf (*see* **Note 8**).
13. Transfer 500 μL of the extract obtained in **step 12** to PVDF mini-uniprep™ (*see* **Note 9**).
14. Add 25 μL of a mixture of internal standards (TPP and DPP) at a concentration of 1 ng/μL (*see* Subheading 2.4), vortex for 10 s (*see* **Note 10**), and close the mini-uniprep™ to filter the extract.
15. The final extract is filtered and then direct analyzed using high-performance liquid chromatography-tandem mass spectrometry (HPLC-MS/MS) with a triple quadrupole (QTRAP+) mass spectrometer utilizing electrospray ionization (ESI) (*see* **Note 11**).

4 Notes

1. For the preparation of stock standards, not less than 10 mg of the pesticide standard should be weighed.
2. The method could be adapted for determination of residues of pesticides in cereals products by adding 20 mL of cold water at the sample weighted in the 50 mL Falcon tube and allowed to stand for 1 h. The use of water, before ACN extraction, improved the extraction efficiency since water allows the

swelling of cereals grain [9]. Also, the shaking time facilitates matrix swelling.

3. ACN is the most used extraction solvent, without coextraction of interferent molecules from the matrix and easily parts from water in the second phase [10].
4. MgSO4 allows the best salting-out of ACN and the best overall recoveries, especially of polar analytes. However, it also contributes to the residual water content in the ACN layer, which helps to control the polarity of the extraction solvents and thus increases the selectivity of extraction [8].
5. Per sample, the mixture of salts for the liquid-liquid partitioning step contains 4 g $MgSO_4$, 1 g NaCl, 1 g of trisodium citrate dihydrate, and 0.5 g of disodium hydrogen citrate sesquihydrate. Per sample, the mixture of salts for cleanup step contains 900 mg of anhydrous $MgSO_4$ and 150 mg of PSA. In order to facilitate the weighing and subsequent transfer of the salt mixture to the Falcon tubes containing the sample and solvents, use glass weighing boats.
6. Shake vigorously by hand after adding the salt mixture, as these can form lumps in the tube, which is more easily broken up using this approach, rather than just vortex shaking.
7. The cleanup step was performed with a d-SPE using anhydrous $MgSO_4$ and PSA. $MgSO_4$ was used to remove the remaining water from the extract while PSA is useful for removing many polar interfering compounds from the extract such as organic acids and sugars [10].
8. The addition of 220 μL of acetonitrile (ACN) reduces matrix-induced effects by ensuring that both sample solutions and calibration standards contain similar concentrations of coextracted matrix components, thereby minimizing analytical variability. This volume adjustment is an essential step to ensure consistency and reliability in pesticide residue analysis, particularly when applied to gas chromatography (GC) analysis.
9. Mini-uniprep™ is a syringeless filter from Whatman™, efficient in filtration and is a fast and cost-effective option. It is designed for the preparation and analysis of samples for HPLC/UHPLC. In this method, it is preferable to use transparent PVDF filters with a 0.2 μm pore size. Their suitability may vary depending on the physicochemical properties of the target analytes. Therefore, it is advisable to ensure compatibility with the specific pesticide residues being analyzed.
10. The ISs are added just before LC-MS analysis to correct for instrumental variations.
11. The chromatographic method is described by Melo et al. (2020) [11].

12. For determination of residues of pesticides in samples, matrix-matched calibration curve should be prepared at 10 concentration levels, i.e., at 5, 10, 20, 30, 40, 50, 60, 70, 80, and 100 μg/kg, using a matrix blank. The matrix blank is a material (a sample, or a portion or extract of a sample) free of the substance(s) to be analyzed. An extraction is performed with the matrix blank as described by 3. Methodology up to **step 12**, and then the matrix-matched calibration curve is prepared as described below. First, a solvent calibration curve is prepared with defined volumes of the mix of internal standards at 2.5 ng/μL and a working solution at 250 μg/kg in ACN (5–200 μg/kg). Then, 500 μL is taken from the levels of the solvent calibration curve, and 500 μL of blank sample and 10 μL of internal standards at 2.5 ng/μL are added to each level (5–100 μg/kg). For instance, for the construction of the 5 μg/kg level, 500 μL is taken from the 10 μg/kg level of the solvent curve and added to 500 μL of a blank sample and 10 μL of a solution of internal standards. Following, each vial is injected into the HPLC-MS/MS.
13. The method could be applied to fruit and their by-products, such as peels and seeds [12, 13].
14. This method has been successfully miniaturized for the determination of pesticide residues in by-product extracts. The miniaturized version reduces the sample amount to 100 mg (compared to the original 10 g), and the extraction salts to 400 mg (originally 6.5 g), consisting of $MgSO_4$, NaCl, sodium citrate, and disodium hydrogen citrate sesquihydrate in a ratio of 4:1:1:0.5 w/w/w/w. The cleanup step was also reduced to 15 mg of PSA and 90 mg of anhydrous $MgSO_4$ (from 105 mg total in the base method). Importantly, this miniaturized protocol maintained the analytical performance, with no significant compromise in sensitivity or recovery, making it a sustainable and efficient alternative when low amounts of sample are available [14].

References

1. FAO Per Capita Consumption of Fruit Available online: https://ourworldindata.org/grapher/fruit-consumption-per-capita?tab=chart&country=~OWID_WRL. Accessed 26 Feb 2025
2. Liu N, Huang J, Liu X, Wu J, Huang M (2024) Pesticide-induced metabolic disruptions in crops: a global perspective at the molecular level. Sci Total Environ 957. https://doi.org/10.1016/j.scitotenv.2024.177665
3. Jankowska M, Kaczyński P, Hrynko I, Rutkowska E, Iwaniuk P, Ilyasova G, Łozowicka B (2024) Dietary risk assessment of children and adults consuming fruit and vegetables with multiple pesticide residues. Chemosphere 369. https://doi.org/10.1016/j.chemosphere.2024.143858
4. Viviani B, Midali M, Serafini MM, Mangas I (2025) Implementing AOPs use for pesticide neurotoxicity assessment. Curr Opin Toxicol 41:100505
5. Srivastava A, Srivastava AK, Pandeya A, Pant AB (2023) Pesticide mediated silent

neurotoxicity and its unmasking: an update on recent Progress. Toxicology 500:153665

6. Asadikaram G, Ashrafi MR, Darvish Moghaddam S, Abolhassani M, Bagheri F (2025) Association of High Organochlorine Pesticide Serum Levels with oxidative stress in intestinal metaplasia and gastric cancer. Heliyon 11. https://doi.org/10.1016/j.heliyon.2024.e41599
7. Anastassiades M, Lehotay SJ, Štajnbaher D, Schenck FJ (2003) Fast and easy multiresidue method employing acetonitrile extraction/partitioning and "dispersive solid-phase extraction" for the determination of pesticide residues in produce. J AOAC Int 86:412–431. https://doi.org/10.1093/jaoac/86.2.412
8. Rejczak T, Tuzimski T (2015) A review of recent developments and trends in the QuEChERS sample preparation approach. Open Chem 13:980–1010. https://doi.org/10.1515/chem-2015-0109
9. Carreiró F, Barros SC, Brites C, Mateus AR, Ramos F, Torres D, Silva AS (2024) Validation of an HPLC-MS/MS method for the quantification of pesticide residues in Rice and assessment of the washing effect. Food Chem X 24: 101938. https://doi.org/10.1016/j.fochx.2024.101938
10. Perestrelo R, Silva P, Porto-Figueira P, Pereira JAM, Silva C, Medina S, Câmara JS (2019) QuEChERS – fundamentals, relevant improvements, applications and future trends. Anal Chim Acta 1070:1–28. https://doi.org/10.1016/j.aca.2019.02.036
11. Melo MG, Carqueijo A, Freitas A, Barbosa J, Silva AS (2020) Modified QuEChERS extraction and HPLC-MS/MS for simultaneous determination of 155 pesticide residues in Rice (Oryza Sativa L.). Foods 9. https://doi.org/10.3390/foods9010018
12. Mateus ARS, Mariño-Cortegoso S, Barros SC, Sendón R, Barbosa L, Pena A, Sanches-Silva A (2024) Citrus by-products: a dual assessment of antioxidant properties and food contaminants towards circular economy. Innovative Food Sci Emerg Technol 95:103737. https://doi.org/10.1016/J.IFSET.2024.103737
13. Mateus ARS, Barros SC, Cortegoso SM, Sendón R, Barbosa-Pereira L, Khwaldia K, Pataro G, Ferrari G, Breniaux M, Ghidossi R et al (2024) Potential of fruit seeds: exploring bioactives and ensuring food safety for sustainable Management of Food Waste. Food Chem X. https://doi.org/10.1016/j.fochx.2024.101718
14. Mateus ARS, Barros SC, Pena A, Sanches-Silva A (2025) Miniaturized QuEChERS extraction (MQuEChERS) combined with HPLC-MS/MS as new analytical method for determination of 105 residues of pesticides in fruit by-products extracts. Food Chem 481. https://doi.org/10.1016/j.foodchem.2025.143898

Chapter 2

Microextraction Method for Analysis of Polybrominated Diphenyl Ethers and Their Methoxylated Congeners in Seafood

Mateus Henrique Petrarca, José Oliveira Fernandes, and Sara Cristina Cunha

Abstract

An accurate, selective, and sensitive method, based on gas chromatography coupled to tandem mass spectrometry (GC-MS/MS), is described for the simultaneous determination of nine polybrominated diphenyl ethers (PBDEs) congeners and eight methoxylated-brominated diphenyl ethers in several seafood matrices, including clams, crabs, fish muscle, macroalgae, and mussels. The analytical protocol stands out with a microextraction step based on salting-out-assisted liquid-liquid partitioning, utilizing a reduced amount of water-miscible organic solvent (acetonitrile) and salt mixture (sodium chloride and magnesium sulfate). Another distinguished feature is the combination of distinct cleanup procedures to minimize matrix effects, considering the complex and heterogeneous seafood matrices. Dispersive solid-phase extraction (d-SPE) with EMR-Lipid® sorbent was particularly applied to remove fats and other non-polar co-extractives, followed by extract treatment with sulfuric acid due to its strong oxidizing nature; then, the cleanup was completed on a lab-made solid-phase extraction (SPE) column based on aluminum oxide neutral sorbent, using a disposable pipette tip as support.

Key words Environmental pollutants, Fish, GC-MS/MS, PBDEs, Persistent organic pollutants, SPE cleanup

1 Introduction

With more than 200 possible congeners, polybrominated diphenyl ethers (PBDEs) comprise a wide group of anthropogenic chemicals industrially used in electronic castings and circuitry, furniture, plastics, textiles, and other ordinary products to improve their resistance to flammability [1]. This class of flame retardants consists of brominated neutral aromatic compounds constituted by two phenyl rings linked by an oxygen atom (Fig. 1), whose congeners differ in the position and number of the bromine atoms in the phenyl rings [1]. In general, the PBDEs exhibit high hydrophobicity and

Paula Alvito and Helena Oliveira (eds.), *Food Contaminants Toxicity*, Methods and Protocols in Food Science,
https://doi.org/10.1007/978-1-0716-5096-7_2,

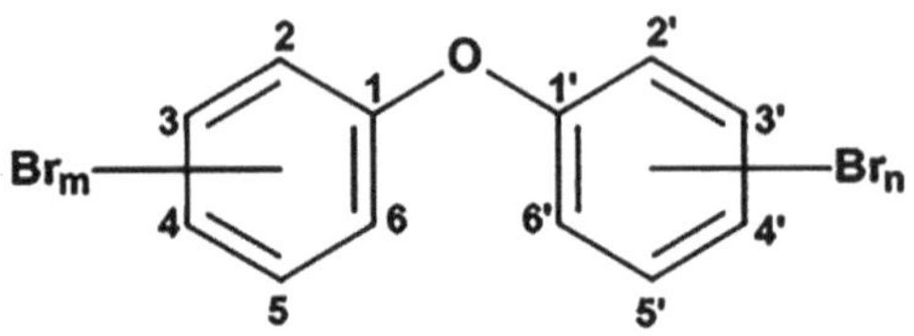

Fig. 1 Basic chemical structure of polybrominated diphenyl ether congeners [1]. The chemical structure was drawn using the Accelrys Draw 4.2 software

low water solubility (octanol-water partition coefficients, KOW, in the range 5.94–12.11), as well as low volatility (vapor pressure in the range 1.64 × 10-12–2.32 × 10–5, kPa at 25 °C) [1]. The PBDEs stand out by their persistence, bioaccumulation capacity in animal and human tissues, biomagnification in trophic chains, and high potential for long-range environmental transport [1, 2]. Specifically, tetra-, penta-, hexa-, hepta-, and deca-bromodiphenyl ethers were included in the Stockholm Convention as persistent organic pollutants (POPs) [2]. Certain countries, such as those in the European Union, have imposed restrictions on the production and use of technical PBDE formulations, including a maximum concentration of 0.1% of PBDEs by weight in homogeneous materials [3].

Generally, PBDEs are not chemically bound to the polymers; hence, they may continually be released from industrial and household products, accumulating in the environment and contaminating food [1, 4, 5]. In the environment, PBDEs can undergo aerobic biotransformation as well as abiotic degradation, including photo-oxidation, photolysis, reductive debromination, and radical reactions, resulting in methoxylated-brominated diphenyl ethers (MeO-BDEs), hydroxylated-brominated diphenyl ethers (OH-BDEs), chlorinated PBDEs, and less-brominated PBDEs [1, 4, 6, 7]. In particular, MeO-BDEs metabolites share structural similarities with their precursors, highlighting a lipophilic nature, bioaccumulative potential, environmental persistence, and toxic effects comparable with the PBDEs [1, 4, 6, 7].

The EFSA Panel on Contaminants in the Food Chain (CONTAM) in the Scientific Opinion on Polybrominated Diphenyl Ethers (PBDEs) in Food identified the main in vivo toxicological effects associated with PBDEs from animal experiments, especially, on the liver, thyroid hormone homeostasis, and reproductive and nervous systems, with effects on neurodevelopment as the critical endpoint [1, 5]. Moreover, PBDEs may induce the generation of reactive oxygen species, leading to DNA damage [1]. Non-occupational human exposure to the ubiquitous PBDE pollutants can occur through diet, inhalation, and ingestion of indoor dust from houses, cars, and office environments [1]. However, the intake of contaminated foods, more precisely seafood, has been recognized as the main exposure route. Specifically, BDE-47, BDE-99, BDE-100, BDE-153, BDE-154, BDE-183, and BDE-209 are among the PBDE congeners more commonly found in seafood, with the highest contents achieving up to 81,800 μg/kg (lipid weight basis) reported for BDE-47 in mud carp species [1, 4, 8].

Sample preparation techniques that comply with the Green Analytical Chemistry (GAC) principle have been prioritized in food analysis in recent years [9]. Accordingly, miniaturization of classical extraction techniques has been an assertive strategy to enhance the greenness of analytical protocols toward environmental friendliness [9, 10]. Significant reduction in the amount of sample, organic solvents, and other hazardous reagents, associated with lower waste generation and reduced cost, undoubtedly comprises the main advantages of miniaturized extraction techniques [11, 12]. Moreover, the straightforwardness and adaptability to several matrices facilitate the coupling of microextraction techniques with each other or conventional sample treatments for analyte isolation and enrichment, cleanup, and preconcentration purposes [12].

In this chapter, a microextraction method is described for the simultaneous analysis of nine PBDEs and eight methoxylated congeners (Table 1) in several seafood matrices of complex composition, using gas chromatography coupled to tandem mass spectrometry (GC-MS/MS). The analytes are extracted using the QuEChERS (Quick, Easy, Cheap, Effective, Rugged, and Safe) approach based on salting-out-assisted liquid-liquid partitioning. The procedure involves a water-miscible organic solvent (acetonitrile) and the addition of a salt (sodium chloride), resulting in a biphasic system [9, 13, 14]. In the salting-out phenomenon, one of the solvents preferentially solvates the electrolyte, making it unavailable to dissolve in the other solvent, thus achieving the phase separation [9, 13]. Moreover, the extensive contact surface benefits the extraction of analytes, facilitating mass transfer kinetics between the two liquid phases [9, 14]. Acetonitrile has been demonstrated to be a suitable organic solvent for the extraction of pollutants of distinct nature [15–18], especially owing to the poor extraction of both highly polar proteins and highly lipophilic components such as fat, waxes, and non-polar pigments [19]. Another advantage of acetonitrile is its ability to separate from water in the sample using a salt mixture without the need for non-polar co-solvents to induce liquid phase separation [19]. In addition to sodium chloride to initiate the salting-out effect, its combination with magnesium sulfate contributes to increasing recoveries of certain analytes. Specifically, magnesium sulfate presents a high ability to bind a large fraction of water, reducing the percentage of water

Table 1
GC-MS/MS parameters for the determination of polybrominated diphenyl ethers (PBDEs) and methoxylated-brominated diphenyl ethers (MeO-BDEs) in seafood

Analytes	Acronym	Molecular formula	Molecular weight (g/mol)	RT (min)[a]	Quantification transition (*m/z*)	CE (kV)	Qualification transition (*m/z*)	CE (kV)	DT (ms)
PBDEs									
2,4,4′-Tribromodiphenyl ether	BDE-28	$C_{12}H_7Br_3O$	406.8954	6.46	407.5→247.9	20	405.6→245.9	20	22
3,4,4′-Tribromodiphenyl ether	BDE-37	$C_{12}H_7Br_3O$	406.8954	6.76	407.5→247.9	27	405.6→245.9	27	22
2,2′,4,4′-Tetrabromodiphenyl ether	BDE-47	$C_{12}H_6Br_4O$	485.7914	7.95	485.7→326.0	20	483.6→323.9	20	12
3,3′,4,4′-Tetrabromodiphenyl ether	BDE-77	$C_{12}H_6Br_4O$	485.7914	8.67	485.7→378.9	27	483.6→376.8	27	12
2,2′,4,4′,5-Pentabromodiphenyl ether	BDE-99	$C_{12}H_5Br_5O$	564.6875	9.78	563.6→403.9	15	403.8→296.9	35	20
5′-Fluoro-3,3′,4,4′,5-pentabromodiphenyl ether	FBDE-126	$C_{12}H_5Br_4FO$	503.7819	10.39	581.7→472.8	25	581.7→421.9	25	15
2,2′,4,4′,5,5′-Hexabromodiphenyl ether	BDE-153	$C_{12}H_4Br_6O$	643.5836	11.84	645.4→483.5	22	643.6→483.4	13	17
2,2′,4,4′,5,6′-Hexabromodiphenyl ether	BDE-154	$C_{12}H_4Br_6O$	643.5836	11.05	645.4→483.5	10	643.6→483.4	13	15
2,2′,3,4,4′,5′,6-Heptabromodiphenyl ether	BDE-183	$C_{12}H_3Br_7O$	722.4796	13.87	723.8→561.6	35	721.6→561.6	35	35
MeO-BDEs									
2′-Methoxy-2,3′,4,5′-tetrabromodiphenyl ether	2′-MeO-BDE-68	$C_{13}H_8Br_4O_2$	515.8174	8.59	515.9→419.8	25	515.9→355.7	25	12

4′-Methoxy-2,2′,4,5′-tetrabromodiphenyl ether	4′-MeO-BDE-49	$C_{13}H_8Br_4O_2$	515.8174	9.35	515.7→500.5	27	515.7→355.9	30	20
4′-Methoxy-2,2′,4,5,5′-pentabromodiphenyl ether	4′-MeO-BDE-101	$C_{13}H_8Br_5O_2$	595.7214	11.27	593.8→433.7	27	593.8→420.6	30	17
4-Methoxy-2,2′,4′,5,6′-pentabromodiphenyl ether	4-MeO-BDE-103	$C_{13}H_8Br_5O_2$	595.7214	10.46	593.6→578.6	20	593.6→433.7	18	15
5-Methoxy-2,2′,4,4′-tetrabromodiphenyl ether	5-MeO-BDE-47	$C_{13}H_8Br_4O_2$	515.8174	9.27	515.6→355.9	22	515.6→340.7	25	20
5′-Methoxy-2,2′,4,4′,5-pentabromodiphenyl ether	5′-MeO-BDE-99	$C_{13}H_8Br_5O_2$	595.7214	11.17	593.6→433.6	20	593.6→418.7	30	17
5′-Methoxy-2,2′,4,4′,6-pentabromodiphenyl ether	5-MeO-BDE-100	$C_{13}H_8Br_5O_2$	595.7214	10.37	593.7→433.8	20	593.7→417.0	20	15
6-Methoxy-2,2′,4,4′-tetrabromodiphenyl ether	6-MeO-BDE-47	$C_{13}H_8Br_4O_2$	515.8174	8.80	515.7→419.8	25	515.7→355.9	25	12
6-Methoxy-2,2′,4,4′-tetrabromo [$^{13}C_{12}$]diphenyl ether	^{13}C-6-MeO-BDE-47	$^{13}C_{12}{}^{12}C_1H_8Br_4O_2$	527.7292	8.80	527.2→433.8	25	527.2→367.9	22	12

CE collision energy, *DT* dwell time, *RT* retention time

[a]Retention time observed using a DB-5MS capillary column (30 m × 0.25 mm × 0.25 μm; Agilent Technologies, USA), according to the temperature program described in Subheading 3.3. *GC-MS/MS analysis*

remaining in the organic phase [19]. Thus, the polarity of the organic phase is decreased, making it more receptive to non-polar compounds like PBDEs. Moreover, the exothermic hydration process of magnesium sulfate contributes to the extraction of non-polar analytes [19].

Considering the complex and heterogeneous matrices of seafood covered in the proposed protocol (clams, crabs, macroalgae, mussels, and fish muscle), distinct cleanup procedures were combined for the purification of extracts, including dispersive solid-phase extraction (d-SPE), followed by treatment with sulfuric acid and then solid-phase extraction (SPE). The d-SPE technique is recognized by its attributes such as (i) low chemical sorbent quantity, typically 10–50 mg per mL of extract; (ii) the possibility of a combination of different sorbents in the same procedure; (iii) no requirement for additional co-solvents; and (iv) easy and simple execution. The technique involves adding a solid sorbent to the sample extract, followed by rapid shaking, and then obtaining the cleaned-up extract through centrifugation. The dispersion of the sorbent as fine particles throughout the sample extract establishes a high contact surface between the phases, effectively providing the retention of matrix co-extractives [20]. Based on the high fat content of certain fish species, EMR-lipid® sorbent has been suitably used in d-SPE technique to remove lipids and other non-polar matrix co-extractives [15–18, 21]. The efficiency of the EMR-lipid® sorbent is associated with size exclusion process and hydrophobic interactions, where long-chain hydrocarbons such as lipids are trapped within the sorbent structure [15, 21]. Nonetheless, the removal of lipids and other organic materials is completed by an additional extract treatment with sulfuric acid, attributed to its strong oxidizing nature [22], with no significant loss of the PBDEs.

The SPE technique stands out for its operational simplicity, limited use of organic solvents, and extensive applicability, including cleanup, class fractionation, extraction, medium exchange, extract purification, and analyte isolation and concentration [23–25]. The SPE process generally involves (i) preparation of SPE column; (ii) conditioning/equilibration of solid-phase material to increase effective surface area and to minimize interferences; (iii) loading of aqueous sample (or organic extract) onto a solid sorbent phase; (iv) washing out of interfering compounds; and then (v) eluting of target analytes from the sorbent by flushing with organic solvents [23–25]. The SPE principle is based on the selective distribution of analytes between the liquid phase (extract or aqueous sample) and the solid phase (immobilized sorbent), where the target compounds often present a high affinity for the sorbent phase. Depending on the physico-chemical properties of the sorbent, the mechanisms involved in the separation process include adsorption, partitioning, ion exchange, or size exclusion [23–

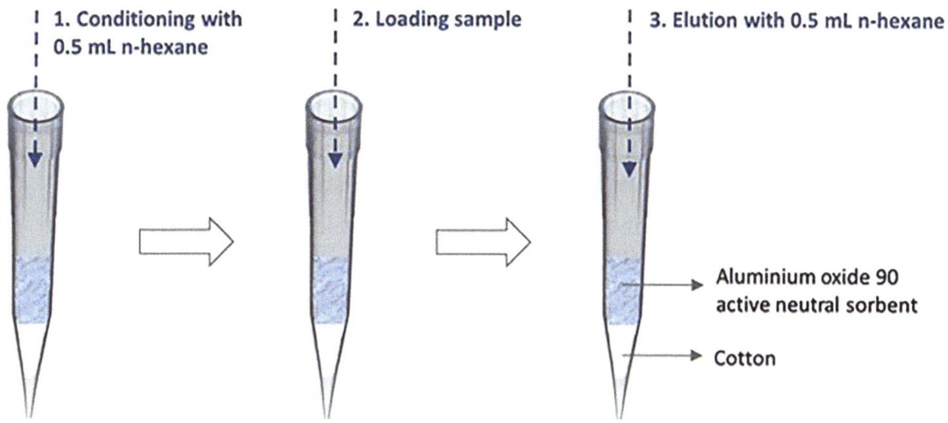

Fig. 2 General scheme of the cleanup procedure using a lab-made SPE column based on a disposable pipette tip

25]. Alumina (aluminum oxide) is one of the most important inorganic oxide adsorbents used for SPE, together with silica gel, Florisil® (synthetic magnesium silicate), and diatomaceous earth [23]. The retention of target analytes is highly dependent on the number and type of functional groups present in their chemical structure. Specifically, compounds containing hydrogen-bonding functional groups, including sulfonic and carboxylic acids, phenol and hydroxyl groups, are strongly retained in the inorganic oxide sorbents [23]. In this analytical protocol, an SPE column based on aluminum oxide 90 active neutral sorbent is easily prepared in the laboratory using a disposable pipette tip as support (Fig. 2) to complete the cleanup process after the oxidation of organic material with sulfuric acid.

In summary, the analytical protocol has been successfully applied for monitoring PBDEs and their methoxylated metabolites in various seafood species, achieving appropriate performance characteristics, such as recoveries within the range 70–120%; precision, expressed as relative standard deviation under repeatability and within-laboratory reproducibility conditions, lower or equal to 20%; linearity in matrix-matched calibration curves; and suitable sensitivity with limits of detection between 0.005 and 0.165 μg/kg, and limits of quantification ranging from 0.015 to 0.655 μg/kg [17, 18, 26].

2 Materials

2.1 Reagents

1. Analytical standards of PBDEs (congeners 28, 37, 47, 77, 99, 153, 154, and 183 and 5′-fluoro-3,3′,4,4′,5-pentabromodiphenyl ether) and MeO-BDEs (congeners 2’-MeO-BDE-68, 6-MeO-BDE-47, 13C-6-MeO-BDE-47, 5-MeO-BDE-47,

4-MeO-BDE-49, 5-MeO-BDE-100, 4-MeO-BDE-103, 5'--MeO-BDE-99, and 4'-MeO-BDE-101) with purity ≥98.0%.

2. Acetone, HPLC grade (≥99.5%).
3. Acetonitrile, HPLC grade (≥99.9%).
4. Aluminum oxide 90 active neutral sorbent.
5. Anhydrous magnesium sulfate, analytical reagent grade (≥99.5%).
6. n-Hexane, HPLC grade (≥99.0%).
7. QuEChERS dSPE EMR-Lipid® sorbent (Agilent Technologies, USA).
8. Sodium chloride, analytical reagent grade (≥99.5%).
9. Sulfuric acid, analytical reagent grade (≥95.0%).
10. Toluene, for GC residue analysis (≥99.9%).
11. Trichloroethylene, for GC residue analysis (≥99.5%).
12. Ultrapure water at 18.2 mΩ/cm.

2.2 Preparation of Standard Solutions

1. Prepare individual stock solutions of PBDEs and MeO-BDEs in n-hexane at 500 μg/mL and store in amber glass flasks at −20 °C. Then, from the individual stock solutions, prepare multi-analyte standard solutions at 50 and 10 μg/mL and store them in amber glass flasks at −20 °C. Working standard solutions at 1, 0.5, and 0.1 μg/mL, or other required concentrations, should be prepared weekly in n-hexane before use and stored at 5 °C.

2.3 Activation of QuEChERS d-SPE EMR-Lipid® Sorbent

1. Weigh 0.2 g of QuEChERS d-SPE EMR-Lipid® sorbent into a 5 mL amber screw-capped glass vial and add 1.0 mL of ultrapure water. Shake the mixture gently by hand and centrifuge it at 2500 rpm for 5 min. Discard the supernatant (*see* **Note 1**).

2.4 In-House Preparation of Solid-Phase Extraction (SPE) Column

1. Place a piece of cotton at the bottom of a 1 mL disposable pipette tip and top it with 0.2 g of aluminum oxide 90 active neutral sorbent (Fig. 2). The SPE columns should be wrapped in aluminum foil and stored in a desiccator before use. The SPE columns were not reused.

2.5 Pretreatment of Anhydrous Magnesium Sulfate

1. Heat the anhydrous magnesium sulfate in a muffle furnace at 500 °C for at least 5 h. Store it in a desiccator before use; it can be utilized for 1 month from preparation.

2.6 Pretreatment of Lab Materials

1. All the glassware, including glass centrifuge tubes, should be heated in a muffle furnace at 300 °C for at least 1 h to minimize the background contamination. Before use, all materials used in the sample preparation are rinsed with acetone.

2.7 Procedural Blanks

1. Procedural blanks, containing all reagents and organic solvents involved in the sample preparation approach, are highly recommended for monitoring background contamination. For each batch of samples, simultaneously prepare at least three procedural blanks (*see* **Note 2**).

3 Methods

3.1 Pretreatment of Seafood Samples

1. The analytical protocol covers seafood of distinct species (*see* **Note 3**). The whole soft body tissue of clams and mussels should be separated from the shell; the fish muscles, without skin or bones, are obtained from fish fillets, and the edible tissues of crabs should be removed from the exoskeleton. It is recommended that all samples (clam, crab, fish muscle, macroalgae, and/or mussel) be submitted to the freeze-drying process. Then, the freeze-dried samples should be preserved at 5 °C until the analysis. Previously to the extraction procedure, the samples should be ground in a stainless-steel blender.

3.2 Extraction and Cleanup

1. Weigh 0.5 g of freeze-dried seafood (clam, crab, fish muscle, macroalgae, or mussel) into a 15 mL amber screw-capped glass flask. It is recommended to have three replicates for each sample.
2. Add an aliquot of internal standard mixture solution (BDE-37, BDE-77, FBDE-126, and 13C-6-MeO-BDE-47) to obtain a final concentration within the linear range of calibration curves (*see* **Note 4**).
3. Add 2.0 mL of ultrapure water and 2.5 mL of acetonitrile: toluene (4:1; v/v).
4. Shake the tube in an orbital shaker incubator at 240 rpm under constant agitation at 25 °C for at least 12 h (*see* **Note 5**).
5. Add 1.0 g of muffled anhydrous magnesium sulfate and 0.25 g sodium chloride and seal the tube well.
6. Shake the tube vigorously by hand, ensuring that the crystalline agglomerates are broken up sufficiently (*see* **Note 6**).
7. Centrifuge the mixture at 2500 rpm for 5 min.
8. Transfer a 1.25 mL aliquot of the upper layer (acetonitrile: toluene) to another amber screw-capped glass flask containing 0.2 g of QuEChERS d-SPE EMR-Lipid® sorbent previously activated with water.
9. Shake the tube with the aid of a vortex and centrifuge the mixture at 2500 rpm for 5 min.

10. Take 1.0 mL of the supernatant and transfer it to an amber screw-capped glass flask. Gently evaporate the aliquot at 40 °C under a nitrogen stream.
11. Reconstitute the residue with 0.5 mL of n-hexane (*see* **Note 7**).
12. Add 0.25 mL of concentrated sulfuric acid to the reconstituted extract (*see* **Note 8**).
13. Shake the tube with the aid of a vortex and centrifuge the mixture at 2500 rpm for 5 min.
14. Collect the upper layer (n-hexane) and transfer it to an amber screw-capped glass flask containing 0.5 mL of ultrapure water.
15. Shake the tube with the aid of a vortex.
16. Centrifuge the mixture at 2500 rpm for 5 min and collect the upper layer (n-hexane).
17. Accommodate the SPE column (in-house prepared according to Subheading 2.3) on a suitable support that allows easy collection of eluent (*see* **Note 9**).
18. Condition the SPE column with 0.5 mL of n-hexane slowly (Fig. 2), and discard the organic solvent.
19. Load the SPE column with the extract.
20. The extract should pass through the sorbent material slowly using the gravity force and then discarded (Fig. 2).
21. Elute the target analytes from the solid phase with 0.5 mL of n-hexane (gravity feed elution).
22. Collect the eluent into an amber screw-capped glass flask.
23. Evaporate the eluent at 40 °C to dryness under a gentle nitrogen stream.
24. Reconstitute the extract with 70.0 μL of trichloroethylene (*see* **Note 7**).
25. Transfer the final extract to a glass insert placed into an autosampler vial for the GC-MS/MS analysis.

3.3 GC-MS/MS Analysis

1. A gas-chromatographic system coupled to a triple quadrupole mass spectrometer (GC-MS/MS) with an electron ionization (EI) source set at 70 eV is operated in selected reaction monitoring (SRM) mode (Table 1).
2. A DB-5MS capillary column (30 m length × 0.25 mm internal diameter × 0.25 μm film thickness; Agilent Technologies, USA) or equivalent, with 60–325 °C temperature limits, is properly recommended.
3. Inject 1.0 μL of the extract (in trichloroethylene) in pulsed splitless mode as follows: injector at 300 °C, injection pulse pressure at 32 psi for 1 min and purge flow to split vent of 50 mL/min (*see* **Note 10**).

4. The oven temperature program should be held as stated: initial oven temperature at 150 °C for 1.5 min, increase the temperature to 250 °C at 40 °C/min, then ramp to 320 °C at 7 °C/min, and keep at this temperature for 3 min, with a total run time of 18.5 min.
5. Ultra-high purity helium (99.999%) is used as carrier gas at a constant 1.3 mL/min flow rate.
6. The temperature of the EI source, transfer line, and first and second quadrupoles should be set at 320, 250, 150, and 150 °C, respectively.

3.4 Quantification and Identification Criteria

1. Matrix-matched calibration curves are highly recommended to compensate for matrix effects and quantify the analytes in the samples.
2. Ideally, matrix-matched calibration curves should be prepared for each batch of samples using a blank representative matrix of the selected seafood, containing at least six concentration levels each, including the method's limit of quantification in the linear range (*see* **Note 11**).
3. At least two SRM transitions, one for quantification and another for qualification, are mandatory for each analyte (Table 1).
4. The ion ratio, calculated as an intensity ratio of the less intense transition to that of the more intense, should be within a tolerance of ±30% compared to the matrix-matched calibration curves [27].
5. A maximum tolerance of 0.1 min on the retention time of target analytes is acceptable, considering those observed in the matrix-matched calibration curves [27].

4 Notes

1. Using a micropipette, remove all the supernatant and discard it. A minimal residual water content is ideally expected in the activated sorbent.
2. Constituted by only organic solvents and chemical reagents, the procedural blanks should contain the same solvents, reagents, and quantities and volumes used for the samples, then processed as a sample, including all steps of the analytical protocol. Procedural blanks should be injected before and after each set of samples to monitor possible background contamination.
3. The analytical protocol has been applied to distinct seafood species such as sea lettuce (*Ulva* sp.), rockweed (*Fucus* sp.),

blue mussel (*Mytilus edulis*), mediterranean mussel (*Mytilus galloprovincialis*), Japanese clam (*Ruditapes philippinarum*), green crab (*Carcinus maenas*), common barbel (*Barbus barbus*), European seabass (*Dicentrarchus labrax*), flounder (*Platichthys flesus*), grey mullet (*Mugil cephalus*), mullet (*Liza aurata*), meagre (*Argyrosomus regius*), and sole (*Solea solea*).

4. An internal standard solution is added to the sample before the extraction procedure to keep track of possible variations occurring during the sample preparation method and GC-MS/MS analysis.
5. A large set of samples can be simultaneously shaken overnight, thus facilitating and saving lab work time.
6. Vigorous vortex agitation for at least 1 min can also be alternatively used.
7. Apply vigorous vortex agitation for an efficient reconstitution of the extract.
8. Use safety glasses and protective gloves to manipulate the concentrated sulfuric acid.
9. A vacuum manifold system is not required for the SPE process. A universal pipette tip box can be easily adapted to support the lab-made SPE columns during the SPE procedure.
10. Pulsed spitless injection mode is preferentially applied because it contributes to low matrix effects and high analytical responses.
11. Matrix-matched calibration curves should be prepared using a blank representative matrix of selected seafood, presenting a chemical composition similar to the analyzed sample, in which the content of target analytes is lower than the method's detection limit. The linear range should properly cover the concentration levels expected in the samples.

References

1. EFSA Panel on Contaminants in the Food Chain – CONTAM (2011) Scientific opinion on Polybrominated diphenyl ethers (PBDEs) in food. EFSA J 9(5):2156
2. Stockholm Convention on Persistent Organic Pollutants (POPs) (2024) https://www.pops.int/TheConvention/ThePOPs/TheNewPOPs/tabid/2511/Default.aspx. Accessed 14 Feb 2025
3. The European Parliament and the Council of the European Union (2011) Directive 2011/65/EU of the European Parliament and of the Council of 8 June 2011 on the restriction of the use of certain hazardous substances in electrical and electronic equipment (recast). Off J Eur Union L174/88–100
4. Cruz R, Cunha SC, Marques A, Casal S (2017) Polybrominated diphenyl ethers and metabolites – an analytical review on seafood occurrence. TrAC Trends Anal Chem 87:129–144
5. Talsness CE (2008) Overview of toxicological aspects of polybrominated diphenyl ethers: a flame retardant additive in several consumer products. Environ Res 108:158–167
6. Xu W, Wang X, Cai Z (2013) Analytical chemistry of the persistent organic pollutants identified in the Stockholm convention: a review. Anal Chim Acta 790:1–13

7. Pan Y, Tsang DCW, Wang Y, Li Y, Yang X (2016) The photodegradation of polybrominated diphenyl ethers (PBDEs) in various environmental matrices: kinetics and mechanisms. Chem Eng J 297:74–96
8. Cruz R, Cunha SC, Casal S (2015) Brominated flame retardants and seafood safety: a review. Environ Int 77:116–131
9. Petrarca MH, Braga PAC, Reyes FGR, Bragotto APA (2022) Exploring miniaturized sample preparation approaches combined with LC-QToF-MS for the analysis of sulfonamide antibiotic residues in meat- and/or egg-based baby foods. Food Chem 366:130587
10. Petrarca MH, Vicente E, Tfouni SAV (2024) Single-run gas chromatography-mass spectrometry method for the analysis of phthalates, polycyclic aromatic hydrocarbons, and pesticide residues in infant formula based on dispersive microextraction techniques. Microchem J 197:109824
11. Ramos L (2012) Critical overview of selected contemporary sample preparation techniques. J Chromatogr A 1221:84–98
12. Moreda-Pineiro J, Moreda-Pineiro A (2015) Recent advances in combining microextraction techniques for sample pre-treatments. TrAC Trends Anal Chem 71:265–274
13. Valente IM, Gonçalves LM, Rodrigues JA (2013) Another glimpse over the salting-out assisted liquid-liquid extraction in acetonitrile/water mixtures. J Chromatogr A 1308: 58–62
14. Koltsakidou A, Zacharis CK, Fytianos K (2015) A validated liquid chromatographic method for the determination of polycyclic aromatic hydrocarbons in honey after homogeneous liquid-liquid extraction using hydrophilic acetonitrile and sodium chloride as mass separating agent. J Chromatogr A 1377:46–54
15. Petrarca MH, Fernandes JO, Marmelo I, Marques A, Cunha SC (2022) Multi-analyte gas chromatography-mass spectrometry method to monitor bisphenols, musk fragrances, ultraviolet filters, and pesticide residues in seafood. J Chromatogr A 1663:462755
16. Petrarca MH, Menezes-Sousa D, Ferreira R, Fernandes JO, Vieira LR, Guilhermino L, Cunha SC (2022) Occurrence and risk assessment of endocrine-disrupting compounds in fish muscle: the case study of the Douro River estuary (north East Atlantic Ocean). Environ Res 215:114236
17. Lestido-Cardama A, Petrarca M, Monteiro C, Ferreira R, Marmelo I, Maulvault AL, Anacleto P, Marques A, Fernandes JO, Cunha SC (2023) Seasonal occurrence and risk assessment of endocrine-disrupting compounds in Tagus estuary biota (NE Atlantic Ocean coast). J Hazard Mater 444:130387
18. Menezes-Sousa D, Cunha SC, Vieira LR, Barboza LGA, Guilhermino L, Alonso MB, Torres JPM, Fernandes JO (2021) Polybrominated diphenyl ethers and their methoxylated congeners in Douro river estuary biota: seasonal occurrence and risk assessment. Sci Total Environ 790:147916
19. Anastassiades M, Lehotay SJ, Stajnbaher D, Schenck FJ (2003) Fast and easy multiresidues method employing acetonitrile extraction/partitioning and "dispersive solid-phase extraction" for the determination of pesticide residues in produce. J AOAC Int 86:412–431
20. Petrarca MH, Perez MAF, Tfouni SAV (2022) Bisphenol A and its structural analogues in infant formulas available in the Brazilian market: optimisation of a UPLC-MS/MS method, occurrence, and dietary exposure assessment. Food Res Int 160:111692
21. Han L, Matarrita J, Sapozhnikova Y, Lehotay SJ (2016) Evaluation of a recent product to remove lipids and other matrix co-extractives in the analysis of pesticide residues and environmental contaminants in foods. J Chromatogr A 1449:17–29
22. Zelinkova Z, Wenzl T (2015) The occurrence of 16 EPA PAHs in food – a review. Polycycl Aromat Comp 35:248–284
23. Poole CF (2003) New trends in solid-phase extraction. TrAC Trends Anal Chem 22:362–373
24. Andrade-Eiroa A, Canle M, Leroy-Cancellieri-V, Cerdà V (2016) Solid-phase extraction of organic compounds: a critical review (Part I). TrAC Trends Anal Chem 80:641–654
25. Buszewski B, Szultka M (2012) Past, present, and future of solid phase extraction: a review. Crit Rev Anal Chem 42:198–213
26. Cruz R, Marmelo I, Monteiro C, Marques A, Casal S, Cunha SC (2020) The occurrence of polybrominated diphenyl ethers and their metabolites in Portuguese river biota. Sci Total Environ 713:136606
27. EU Reference Laboratories for Residues of Pesticides (2024) Analytical quality control and method validation procedures for pesticide residues analysis in food and feed SANTE/11312/2021 v2. https://eurl-pesticides.eu/docs/public/tmplt_article.asp?CntID=727. Accessed 14 Feb 2025

Chapter 3

Analysis of Thermic Heterocyclic Aromatic Amines in Meat Products Using One-Step SPE Followed by LC-MS/MS

Thayná F. Zanin, Maria V. Vicente, Fernando G. Tonin, and Alessandra V. Jager

Abstract

The method described employs solid-phase extraction with strong cationic exchange resin with subsequent identification and quantification by LC-MS/MS to determine six heterocyclic aromatic amines in beef and chicken meat: IQ, IQx, MeIQ, MeIQx, 4,8-DiMeIQx, and PhIP.

Key words Heterocyclic aromatic amines, Solid-phase extraction, Liquid chromatography, Mass spectrometry, Beef, Chicken

1 Introduction

Heterocyclic aromatic amines (HAA) are compounds that contain two to five condensed aromatic rings in their structure with one or more nitrogen atoms in their ring [1]. Currently, over 25 HAA have been identified in cooked foods. These are generally divided into two main classes: aminocarbolines (also known as pyrolytic or non-polar HAA) and aminoimidazoazarenes (AIAs, also known as thermic or polar HAA). Cooking procedure, cooking time, type of meat, pH, lipid content, concentration of free amino acids, sugars, creatin(in)e, and the presence of antioxidants are all factors that influence HAA formation. AIAs are predominantly formed at temperatures between 100 °C and 300 °C, which correspond to the conditions typically used in everyday cooking methods. AIAs have a 2-amino-imidazo group and a methyl group linked to one of the nitrogens in the imidazole ring; among the representatives of this class are 2-amino-3-methylimidazo [4,5-f] quinoline (IQ), 2-amino-3-methylimidazo [4,5-f] quinoxylanine (IQx), 2-amino-3,4-dimethylimidazo [4,5-f] quinoline (MeIQ), 2-amino-3,8-dimethylimidazo [4,5-f] quinoxylanine (MeIQx), 2-amino-3,4,8-

Paula Alvito and Helena Oliveira (eds.), *Food Contaminants Toxicity*, Methods and Protocols in Food Science,
https://doi.org/10.1007/978-1-0716-5096-7_3,

trimethylimidazo [4,5-f]quinoxaline (4,8– DiMeIQx), and 2-amino-1methyl-6-phenylimidazo [4,5- b] pyridine (PhIP). The International Agency for Research on Cancer (IARC) has classified IQ as a probable human carcinogen (Group 2A), and MeIQx, MeIQ, and PhIP as possible human carcinogens (Group 2B) [2]. Several studies have already demonstrated the high incidence of HAA in heat-treated meat commonly consumed daily in various countries [3–5]. Therefore, analytical methods for determining these compounds in complex food matrices are becoming increasingly important. Chromatographic separation methods, which are necessarily preceded by a sample preparation and cleanup step, with subsequent quantification by appropriate detection systems, have been used for these analyses [6–8]. The most promising analytical methods currently employ solid phase extraction for sample preparation and determination by liquid chromatography coupled with mass spectrometry (LC-MS/MS) [9–11]. This chapter describes an analytical sample-preparation procedure, adapted from Yan et al. [8], that involves extraction with hydrochloric acid and a one-step solid-phase extraction using a strong cation exchange cartridge, followed by separation and quantification of the six most common AIAs by LC-MS/MS [8]. The procedure has been used to determine IQ, IQx, MeIQ, MeIQ, 4,8- DiMeIQx, and PhIP in beef and chicken fillets.

2 Materials

2.1 Standard Solutions

1. Standard stock solutions: 400 μg.mL^{-1} IQ, IQx, MeIQ, MeIQx, 4,8-DiMeIQx, and PhIP (Toronto Research Chemicals, North York, Ontario, Canada). Weigh around 2.0 mg, separately, in 5.0 mL volumetric flaks and complete to volume with methanol. Shake until completely dissolved (*see* **Notes 1–4**). Store in amber glass flasks at −20 °C until use.
2. Intermediate standard solutions: 200 μg.mL^{-1} and 20 μg.mL^{-1}. Appropriately dilute the stock solution 400 μg.mL^{-1} in methanol using volumetric flasks. Store in amber glass flasks at −20 °C until use.
3. Calibration curve standards: dilution from intermediate standard solutions. Dilute intermediate stock solutions (200 μg.mL^{-1} or 20 μg.mL^{-1}) to obtain the final concentrations of 1.0, 5.0, 10.0, 25.0, 50.0, 100.0, and 200.0 ng.mL^{-1} for IQx, IQ, and MeIQ. For MeIQx dilute to obtain the final concentrations of 1.4, 5.0, 10.0, 25.0, 50.0, 100.0, and 200.0 ng.mL^{-1}, and for 4,8 DiMeIQx 0.5, 5.0, 10.0, 25.0, 50.0, 100.0, and 200.0 ng.mL^{-1}. Finally, dilute to 1.3, 5.0, 10.0, 25.0, 50.0, 100.0, and 200.0 ng.mL^{-1} for PhIP (*see* **Note 5**).

2.2 Solid-Phase Extraction

1. Conical bottom propylene tubes (50 mL).
2. Conical bottom glass tubes (15 mL).
3. Handheld homogenizer.
4. Ultrasonic bath.
5. Centrifugal vacuum concentrator.
6. pH meter.
7. Solid-phase column: Strata™ X-C (200 mg, 6 mL Phenomenex) (*see* **Note 6**).
8. Hydrochloric acid: 0.1 $mol.L^{-1}$ in water.
9. Trichloroacetic acid: 50% (m/v) in water. Weigh 50 g of acid and add 50 mL of ultrapure water. Homogenize until the solid is completely dissolved.
10. Sodium hydroxide: 4 $mol.L^{-1}$ in water. Weigh 16 g of sodium hydroxide pellets and fill 100 mL in a volumetric flask.
11. Elution solution: methanol:ammonia (95:5, v/v). Mix 95 mL of methanol with 5 mL of ammonia solution ($\geq$ 25%) (*see* **Note 7**).

2.3 Chromatography System and Detector

1. Separation column: BEH C18 (2.1 x 150 mm, 1.7 μm, Waters™).
2. Sampler: Acquity FTN (Waters™).
3. LC-MS/MS system: UPLC Acquity I-Class and Xevo TQS (Waters™).
4. Software: MassLynx™ (version 4.1).
5. Mobile phase composition: (A) 10 $mmol.L^{-1}$ ammonium acetate in water (pH 6.8) and (B) 10 $mmol.L^{-1}$ ammonium acetate in methanol:acetonitrile (1:1, v/v) (*see* **Note 8**).

3 Methods

1. This chapter describes an analytical sample preparation procedure involving extraction with hydrochloric acid and a one-step solid-phase extraction using a strong cation exchange cartridge, followed by separation and quantification of the six most common AIAs by LC-MS/MS, adapted from Yan et al. [9].

3.1 Sample Extraction and Cleanup

1. Perform the steps described in Fig. 1. The samples must be prepared in three replicates and each replicate must be injected in triplicate. The mean values were used for the final expression of the results.

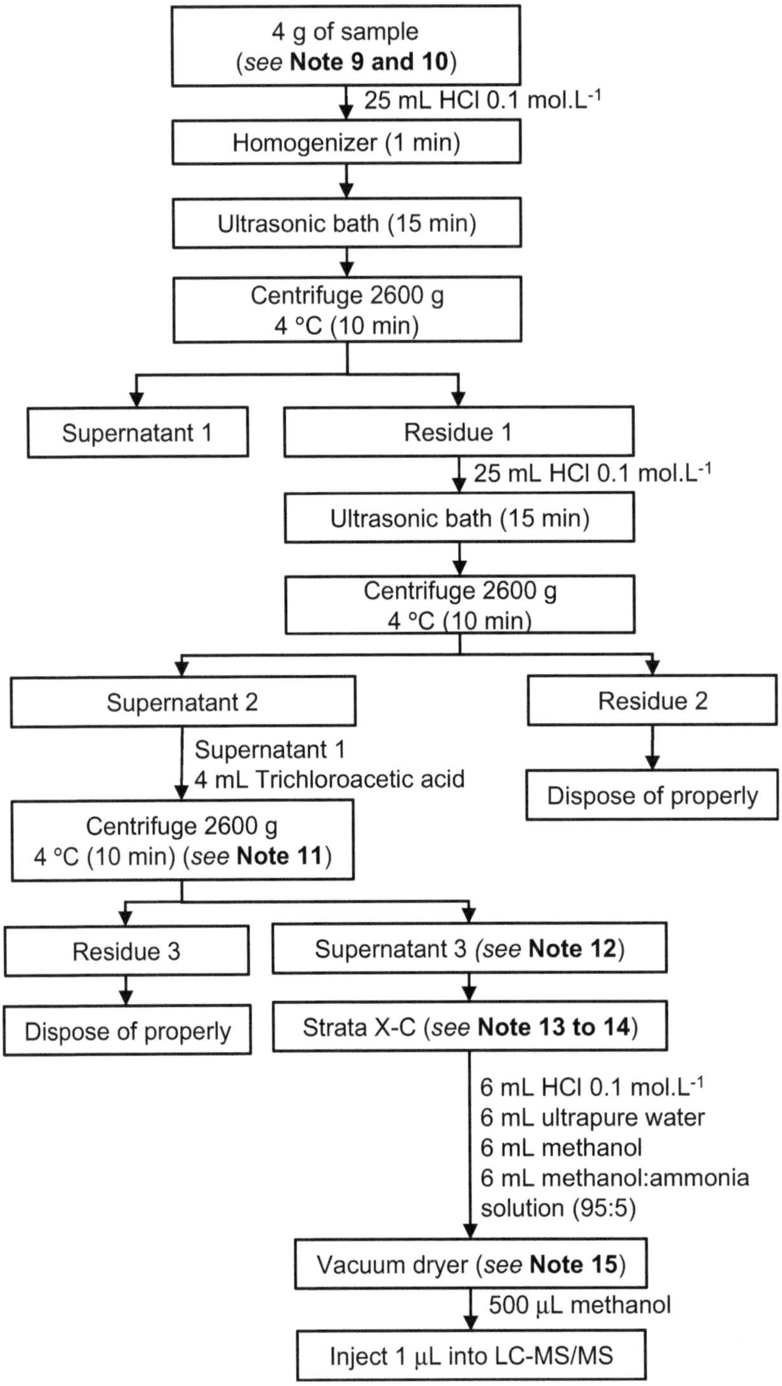

Fig. 1 Flowchart for sample extraction and cleanup

Table 1
MRM transitions, cone voltage, and collision energy used for the quantification and confirmation of aromatic amines by mass spectrometry

Analyte	Precursor ion (*m/z*)	Fragment ions (*m/z*)	Cone voltage (V)	Collision energy (V)
IQ	199	184[a] 157[b]	56 56	24 32
IQx	200	185[a] 159[b]	48 48	24 22
MeIQ	213	198[a] 170[b]	46 46	24 38
MeIQx	214	199[a] 173[b]	50 50	26 20
4,8 DiMeIQx	228	213[a] 187[b]	48 48	24 20
PhIP	225	210[a] 140[b]	54 54	28 38

[a]Quantification transition
[b]Confirmation transition
m/z mass/charge ratio of the precursor ion and its fragments

3.2 Chromatographic Analysis

1. Column temperature: 40 °C.
2. Sampler temperature: 15 °C (*see* **Note 16**).
3. For chromatographic elution, start the gradient at 80% A and hold isocratic for 1.5 min. Decrease linearly to 40% A in 3.5 min and decrease to 0% A in another 1.0 min. Hold at 0% A for 1.0 min and increase to 80% A in 1 min. Equilibrate to initial conditions for another 2 min. Set the mobile phase flow rate at 0.45 mL.min^{-1}.
4. Sample injection: inject 1 μL of samples and standards into the chromatographic system.
5. Mass detector parameters: capillary voltage of 0.60 kV, source temperature of 150 °C, desolvation temperature of 500 °C, desolvation gas at 650 L.h^{-1}, and cone gas flows at 150 L.h^{-1} (*see* **Note 17**).
6. Precursor and fragment ions: set as described in Table 1 (*see* **Note 18**).
7. Quantification: Use the peak area of standard solutions and sample extracts (*see* **Notes 19–21**).

4 Notes

1. All solutions must be prepared with ultrapure water or chromatographic grade solvents.
2. Standard solutions should be prepared by weighing the standard in calibrated and properly maintained analytical balances, and the use of volumetric flasks decontaminated with acetonitrile is highly recommended. For standard storage, it is highly recommended to use new glass vials.
3. Avoid cleaning glassware with detergent when using mass spectrometers. If extremely necessary, make sure to rinse abundantly with ultrapure water and acetonitrile. Ultrasound bath is also helpful.
4. Standard solutions have been shown to be stable over a 6-month period. An extended period may be feasible, but it has not been tested. It is advisable to separate the aliquots from the total volume of the standards to remove from the freezer only the portion that will be used and not return it to the freezer again. Temperature change cycles (room temperature-freezer temperature) can alter the concentration of stock solutions, especially since they are all prepared in methanol, which is quite volatile.
5. The lowest concentration of the solutions corresponds to the limit of quantification of the method for each analyte.
6. Strata X-C columns were used in these experiments. Similar columns are commercially available from other manufacturers. The column technology used for cleanup can particularly affect the recovery and matrix effect.
7. It is advisable to prepare it on the day of use because ammonia is highly volatile and can alter the composition of the solution and consequently influence the elution of amines.
8. It is not necessary to adjust the pH of the aqueous solution to reach a value of 6.8. When prepared correctly, the ammonium acetate solution has this pH. If the pH changes after preparation, it is advisable to check the preparation, the ammonium acetate salt, or the water used in the preparation.
9. Samples should be crushed and frozen if not analyzed immediately. If frozen, allow to reach room temperature to begin the process.
10. Perform all extraction steps in 1 day. Avoid leaving extraction solutions or waiting too long for the next step as this can degrade the analytes.

11. This step is important in precipitating proteins and providing a clean extract for elution by the Strata-X-C column. It is important to ensure that the extract is clear after centrifugation. If not, adjust the centrifugation time. A clear extract is also important to prevent clogging of the column. If this occurs, the entire process that is carried out will be lost.
12. Before eluting the extract through the column, adjust the pH of the extract to 3.0 with sodium hydroxide solution. This step is crucial because the cartridge is a column of exchange of cations. The acidic pH will protonate the amino groups, and thus the analytes will be retained in the column.
13. Furthermore, before extract eluting, the column should be preconditioned according to the manufacturer's instructions: 6 mL of methanol, 6 mL of ultrapure water, and 6 mL of hydrochloric acid. This step should be performed shortly before extraction elution. Do not allow the column to dry out, as this may negatively influence amine extraction.
14. For the entire cleaning process to be truly effective, it is extremely important to observe the elution speed. Keep the flow of solvents through the column at 1 drop per second, never more than that.
15. Collect the eluate in 15 mL glass conical tubes. During vacuum drying, stop drying when the volume has evaporated by about 50%. Wash the walls of the tube with small portions of methanol so that any analyte that has adhered to the walls is recovered in the final solution.
16. Keep the autosampler temperature at a maximum of 15 °C to avoid evaporation of solvents from samples and standards. Autosamplers allow the analysis to be programmed for a long period without the intervention of a technician or student. This prolonged stay in the autosampler, if the temperature is not controlled, can cause preconcentration of all solutions, generating low precision and accuracy of the results.
17. Parameter optimizations should be performed for the equipment that will be used, and the values presented here are for guidance only. This optimization can be performed relatively quickly using the software that comes with modern instruments.
18. The precursor and fragment ions were optimized by direct injection of a 10 $\mu g.mL^{-1}$ standard solution combined with the mobile phase composed of 50% A and 50% B. These parameters should also be checked on the equipment that will be used to verify possible variations.

Table 2
Limit of detection, limit of quantification, linearity, matrix effect, recovery, and precision

HAA	LOD[a] ng.g^{-1}	LOQ[b] ng.g^{-1}	Linearity[c] *r*	Matrix effect[d]	Recovery[e] (%)	Precision (CV, %)
IQx	0.04 (0.32)	0,12 (1,0)	0,9961	1.03	77–88	3.8–5.5
IQ	0.04 (0.32)	0,12 (1,0)	0,9954	0.97	80–92	4.1–8.3
MeIQx	0,05 (0,40)	0,18 (1,4)	0,9974	0.89	86–94	4.6–8.6
MeIQ	0,04 (0,32)	0,12 (1,0)	0,9986	1.05	93–102	4.1–5.6
4,8 DiMeIQx	0,02 (0,16)	0,06 (0,50)	0,9951	1.08	91–99	6.3–8.8
PhIP	0,05 (0,38)	0,16 (1,3)	0,9998	0.93	98–107	5.5–8.4

[a]LOD: Limit of detection in the initial sample. The resulting concentration values in the final injected extract in ng.mL-1 are given in parentheses. Sample preparation concentrates the analytes by a factor of 8
[b]LOQ: Limit of quantification in the initial sample. The resulting concentration values in the final injected extract in ng.mL-1 are given in parentheses. Sample preparation concentrates the analytes by a factor of 8
[c]Correlation coefficient (r)
[d]The effects of the matrix are expressed as the ratio of the slopes of the calibration curves with spiked extracts to the slopes of the calibration curves in the standard solution (methanol). Values > 1.0 indicate increased ionization and < 1.0 indicate suppression of ionization
[e]Recovery, $n = 3$. Recovery = [(C determined – C Initial) / (C Added)] × 100
[f]Coefficient of variation (CV), $n = 3$

19. External standard calibration curves were used to quantify each HAA. The calibration curve should be plotted with the peak areas as a function of the standard solution concentration. The MassLynx software, or any other software used, automatically fits the curve and calculates the concentrations of the sample extract solutions.
20. To express the results of the amine concentrations in the samples in ng, it is necessary to divide the value obtained by 8 (concentration factor) because the whole process started with 4 grams of sample.
21. Table 2 presents figures of merit for the analytical method. The precision of the analytical method was evaluated by calculating the coefficient of variation (CV, %) of three replicates of extractions of spiked samples at three concentration levels: level 1 (LOQ for each AHA), level 2 (10.0 ng.g^{-1}), and level 3 (25.0 ng.g^{-1}). Figure 2 shows the MRM chromatograms for HAA.

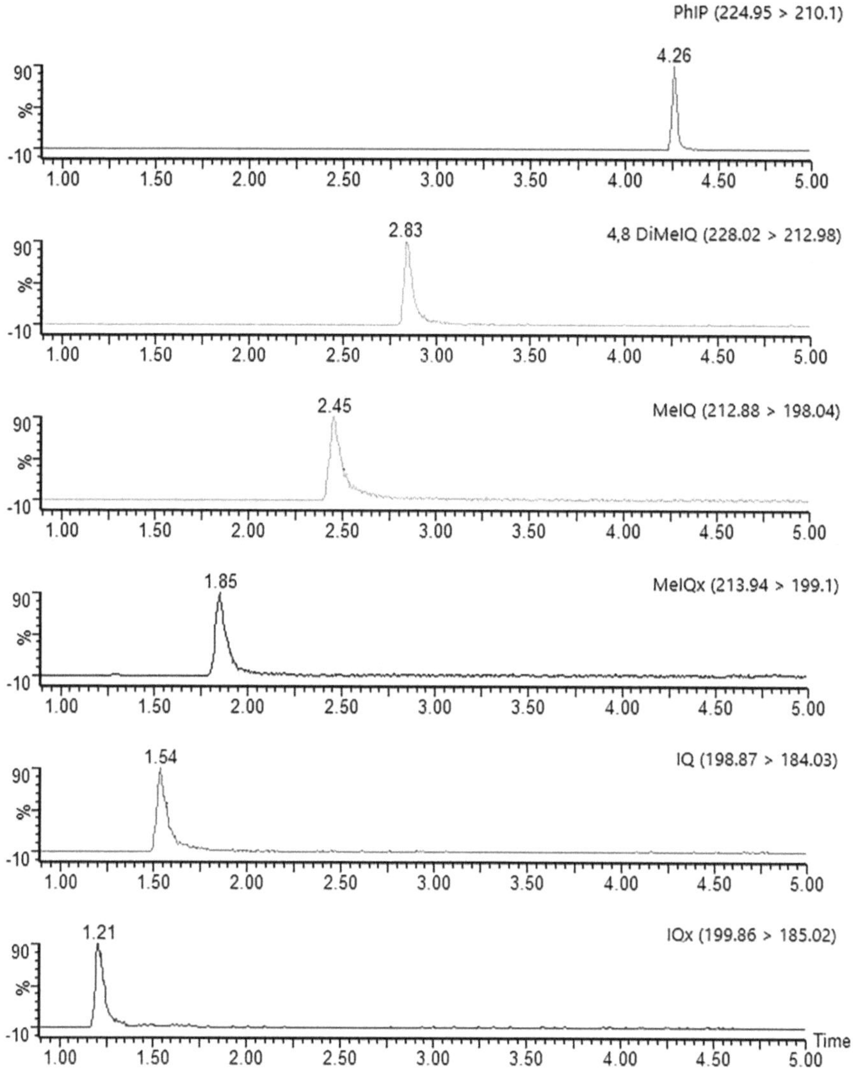

Fig. 2 Chromatogram of the quantification transitions of HAA in solvent (methanol). IQx (1.0 ng.mL^{-1}), IQ (1.0 ng.mL^{-1}), MeIQx (1.4 ng.mL^{-1}), MeIQ (1.0 ng.mL^{-1}), 4.8 DiMeIQx (0.5 ng.mL^{-1}), PhIP (1.3 ng.mL^{-1})

Acknowledgments

This study was financed, in part, by the São Paulo Research Foundation (FAPESP), Brazil—Process Number #2016/09794-8 and Process Number #2018/02911-4#. This study was also financed in part by the Coordenação de Aperfeiçoamento de Pessoal de Nível Superior, Brazil (CAPES)—Finance Code 001. The authors thank Professor Carlos Augusto Fernandes de Oliveira for kindly providing the LC-MS/MS equipment and facilities.

References

1. Barzegar F, Kamankesh M, Mohammadi A (2019) Heterocyclic aromatic amines in cooked food: a review on formation, health risk-toxicology and their analytical techniques. Food Chem 280:240–254
2. IARC Working Group on the Evaluation of Carcinogenic Risks to Humans, International Agency for Research on Cancer, & World Health Organization (1993) Some naturally occurring substances: food items and constituents, heterocyclic aromatic amines and mycotoxins, vol 56. International Agency for Research on Cancer
3. Niklas AA, Pedersen M, Christensen T, Duedahl-Olesen L (2023) Simultaneous determination of heterocyclic aromatic amines and N-nitrosamines in fried bacon cubes and slices using LC-(ESI/APCI)-MS/MS. Food Addit Contam Part A Chem Anal Control Expo Risk Assess 40:493–507
4. Demirok Soncu E, Haskaraca G, Kolsarıcı N (2018) Presence of acrylamide and heterocyclic aromatic amines in breaded chicken meat products and dietary exposure of Turkish population from Ankara based on the food frequency questionnaire study. Eur Food Res Technol 244:501
5. Zhang L, Wang L, Li Y, Xia Y, Chang CM, Xia B, Sosnoff CS, Pine BN, Rey de Castro B, Blount BC (2020) Evaluation of tobacco smoke and diet as sources of exposure to two heterocyclic aromatic amines for the US population: NHANES 2013–2014. Cancer Epidemiol Biomarkers Prev 29:103–111
6. Alaejos MS, Ayala JH, González V, Afonso AM (2008) Analytical methods applied to the determination of heterocyclic aromatic amines in foods. J Chromatogr B 862:15–42
7. Santos FJ, Barceló-Barrachina E, Toribio F, Puignou L, Galceran MT, Persson E, Skog K, Messner C, Murkovic M, Nabinger U, Ristic A (2004) Analysis of heterocyclic amines in food products: interlaboratory studies. J Chromatogr B 802:69–78
8. Yan Y, Zeng MM, Zheng ZP, He ZY, Tao GJ, Zhang S, Chen J (2014) A novel one-step extraction method for simultaneously determining eleven polar heterocyclic aromatic amines in meat products by UHPLC-MS/MS. Anal Methods 6:6437–6444
9. de Moura Neves-Gonçalves T, Pinto E, Viegas O, Braga ARC, de Souza Mesquita LM, Oliveira IMPV, García-Jares C, De Rosso VV, Domene SMÁ (2025) Evaluation of the potential of annatto seed powder to reduce the formation of heterocyclic amines in charcoal-grilled and pan-fried beef patties. Food Chem 462:141015
10. Yoon H, Jung MY, Choi SH, Chun SH, Lee J, Moon B, Lee KW (2024) Effect of air-frying on formation and correlations of polycyclic aromatic hydrocarbons, acrylamide, and heterocyclic aromatic amines in foods and risk assessment. Food Sci Biotechnol 33:2399–2415
11. Hsiao HY, Chen BH, Kao TH (2017) Analysis of heterocyclic amines in meat by the quick, easy, cheap, effective, rugged, and safe method coupled with LC-DAD-MS-MS. J Agric Food Chem 65:9360–9368

Chapter 4

ICP-MS Analysis of Inorganic Contaminants (As, Cd, Pb) in Foodstuffs Preceded by Closed Vessel Microwave Acid Digestion

Marta Ventura, Andreia Rego, Inês Delgado, Sandra Gueifão, and Inês Coelho

Abstract

The present protocol describes the method for the determination of arsenic, cadmium, and lead in foodstuffs using inductively coupled plasma mass spectrometry (ICP-MS) following closed vessel microwave-assisted digestion. This protocol has been optimized and validated for the following food matrices: fish, fruit, and fruit products, Grains and grain-based products, milk and dairy products, and vegetables and vegetable products.

Key words ICP-MS, Inorganic contaminants, Foodstuffs, Microwave digestion

1 Introduction

Inductively coupled plasma mass spectrometry (ICP-MS) is a highly sensitive and well-established analytical technique for multi-element determination at trace and ultra-trace levels in diverse matrices, including food. It is widely recognized for its low detection limits, wide linear dynamic range, high sensitivity, precision, ability to perform multi-element analysis, high sample throughput, and ability to discriminate between isotopes [1]. Compared to inductively coupled plasma optical emission spectroscopy (ICP-OES), ICP-MS offers superior sensitivity and lower detection limits, making it particularly suitable for the determination of trace levels of toxic elements such as arsenic (As), cadmium (Cd), and lead (Pb) [2]. The present method employs an in-house protocol based on the EN 15763 standard [3], optimized and validated for the determination of these elements in various food matrices, including fish and fish products, fruit and fruit products, grains

Paula Alvito and Helena Oliveira (eds.), *Food Contaminants Toxicity*, Methods and Protocols in Food Science,
https://doi.org/10.1007/978-1-0716-5096-7_4,

and grain-based products, milk and dairy products, and vegetables and vegetable products.

The methodology involves two main steps: closed vessel microwave-assisted digestion and analytical determination by ICP-MS. During the digestion step, food samples undergo mineralization in a closed vessel using microwave-assisted acid digestion under controlled conditions, ensuring complete solubilization of target elements while minimizing contaminations and losses. This approach offers high reproducibility, efficient matrix decomposition, rapid and uniform heating, controlled digestion conditions, and rapid processing with improved energy efficiency [4].

Following digestion and dilution, the sample is introduced into the ICP-MS system through a peristaltic pump, ensuring a continuous and stable flow. The sample is then converted into an aerosol by a nebulizer and carried into the argon plasma, where high temperatures (~7500 K) facilitate desolvation, vaporization, atomization, and ionization. The resulting ions pass through a vacuum interface and a lens system that directs them to a quadrupole mass analyzer, where they are separated based on their mass-to-charge ratio. Finally, detection and quantification are performed using an electron multiplier, with external calibration achieved through the use of appropriate standard solutions [2].

Despite its advantages, ICP-MS is susceptible to spectral interferences, including isobaric overlaps, doubly charged ions, and polyatomic species. To mitigate these issues, modern instruments are equipped with collision/reaction cells (CRC), which enhance accuracy by reducing or eliminating interferences [1, 2].

Through the introduction of a gas or gas mixture, the collision cell technology (CCT) and the dynamic reaction cell (DRC) enable interaction with the ion beam, resulting in the chemical removal of polyatomic interferences via collisional dissociation and targeted ion-molecule reactions [5].

Using CCT allows the gas to collide with polyatomic interference, causing it to lose energy. Since polyatomic species are larger, they undergo more collisions and lose more energy than analytes do. The lower-energy interference is then separated from the higher-energy analyte through a process designated as energy discrimination. The system facilitates both reactive and inert collisions, effectively minimizing spectral interferences while enhancing ion focusing and thermalization [1]. DRC uses specific reaction gases that interact chemically with interfering ions or analyte ions to remove known, reactive interferences from each analyte isotope [1].

Alternatively, correction equations may be used to correct for spectral interferences. These are mathematical equations that adjust the signal of the analyte of interest according to the signal of specific interferents, reflecting an existing correlation between both signals [6]. In general, manufacturers provide tables with correction equations for their instruments. However, it is recommended that laboratories verify the applicability of such equations by validating them or establishing new ones if necessary. The approach followed in this work involves the use of correction equations.

The contaminants targeted in the present protocol (As, Cd, and Pb) are among the top ten major public health concern chemicals as defined by the Agency for Toxic Substances and Disease Registry [7], and their exposure mainly occurs through contaminated air, water, and food. The European Union (EU) has established regulatory limits on inorganic arsenic in rice and rice-based products, as rice tends to absorb higher levels of As from water and soil [8]. Monitoring the total amount of As in rice may be sufficient when concentrations do not exceed the limits established for inorganic As. The EU has also set maximum levels for cadmium in fruits, vegetables, fish, and cereal products, as well as for lead in fruits, vegetables, and fish products, to minimize dietary exposure and protect public health [9].

The present protocol was specifically designed for use with the Thermo X Series II ICP-MS and the Milestone Ethos 1 Microwave Digestion System; however, it can be adapted for use with similar equipment. By integrating microwave-assisted digestion with ICP-MS, this method ensures accurate and reliable determination of As, Cd, and Pb in foodstuffs, meeting regulatory and safety requirements while providing a robust tool for food safety monitoring.

2 Materials

2.1 Materials (See Note 1)

1. Beakers made of perfluoroalkoxy alkane (PFA) or polypropylene (PP) of various volumes.
2. Centrifuge tubes, 15 and 50 mL, made of PP.
3. Funnels, made of PFA or PP.
4. In-line internal standard addition system.
5. Measuring cylinders made of PFA or PP of various volumes.
6. PFA flasks of various volumes.
7. PFA or PP volumetric flasks of various volumes.

2.2 Equipment

1. Acid distiller
2. Analytical balance with a minimum resolution of 0.0001 g.
3. Calibrated fixed-volume and variable-volume micropipettes, along with corresponding tips.
4. Freezer (≤ -18°C).
5. Fume cupboard.
6. ICP-MS: Thermo X series II or equivalent.

7. Microwave digestion system Milestone Ethos 1 or equivalent.
8. Refrigerator (5 ± 3°C).
9. Water purification system.

2.3 Gases

1. Argon with a purity of more than 99.996% and a water content of less than 5 mg kg^{-1}.

2.4 Reagents and Standards

Prepare all solutions using high analytical-grade reagents in a clean room.

1. Blank: The blank solution contains water (H_2O) and the same amount of nitric acid (HNO_3) and internal standard as the calibration solutions.
2. Hydrogen peroxide (H_2O_2): Use H_2O_2 with a concentration of 30%, specific for trace element analysis.
3. Internal standard solution: A commercially available monoelemental internal standard solution should be used. One or more elements may be chosen as internal standards. Suitable elements include germanium (Ge), yttrium (Y), and indium (In). Other elements may be used if appropriate. All prepared solutions (samples, standards, blanks, quality controls, etc.) must contain the same concentration of internal standards (*see* **Notes 2–4**).
4. HNO_3: Pro analysis (65% v/v), after purification in the laboratory using an acid distillation system or commercially available suprapure HNO_3 (65%).
5. H_2O: Use ultrapure H_2O, Type I, obtained using a water purification system.
6. Optimization solution, 10 μg L^{-1}: Multielement tuning solution with a minimum of 10 elements, including lithium, indium, and uranium, covering all the mass range. A daily performance report should be conducted to assess the stability and sensitivity of the equipment, mass accuracy, and the presence of oxides and doubly charged ions.
7. Reference material: Wherever possible, reference materials of a similar matrix and with certified concentrations of the elements of interest should be used.
8. Standard solutions for calibration and internal quality control: Calibration and quality control standards should be prepared using high-purity multi-element or single-element ICP-MS stock solutions (*see* **Note 5**).

2.5 Material Decontamination Before Usage

1. All resistant materials (PFA, PTFE, quartz) are individually decontaminated in a 10% (v/v) HNO_3 solution for at least 8 h.

2. All disposable materials made of PP, such as centrifuge tubes, should be individually decontaminated in a 5% (v/v) HNO_3 solution for at least 8 h.
3. All materials should be rinsed with ultrapure water after decontamination.
4. To wash the microwave vessels, perform a blank digestion using a mixture of 5 mL HNO_3 and 5 mL H_2O, followed by rinsing with ultrapure water (*see* **Note 6**).

3 Methods

3.1 Sample Preparation

1. The sample preparation should be performed in a clean room. Before microwave digestion, samples should be thoroughly homogenized to ensure that each subsample accurately represents the original sample. The utensils used should preferably be made of plastic or ceramic, as contact between the sample and metal objects should be avoided to prevent contamination. Exclusively, titanium blades should be utilized in the homogenizer.

3.2 Sample Storage

1. The samples must be kept under storage conditions that maintain their original composition. Depending on the matrix and the intended storage period, samples can be stored:
   - Refrigerated or frozen
   - Freeze-dried (in which case the moisture content must be determined)
   - Stored in a vacuum
   - Stored in the original packaging

3.3 Microwave Digestion

Each digestion cycle should include at least one blank (a vessel containing all reagents but no sample) and at least one quality control sample (either reference material or a fortified sample). The reference vessel, which uses a probe to monitor the pressure and temperature of the microwave digestion, should contain the sample that is most difficult to mineralize, never a blank (*see* **Notes 7** and **8**).

1. Place the PTFE microwave vessel on the balance, tare, and weigh between 0.1 and 1 g, depending on the matrix.
2. Weigh at least two replicates of each sample.
3. In a fume cupboard, add 4 mL of ultrapure HNO3, 1 mL of H2O2, and 3 mL of ultrapure H_2O to the vessel and close it (*see* **Note 9**).

Table 1
Digestion program for 12 vessels

Step	Power (w)	Temperature (°C)	Time (min)
1	1200	90	25
2	1200	90	5
3	1200	180	15
4	0	180	5
5	1200	210	5
6	1200	210	12
7	0	210	5
8	550	90	6

4. Place the vessels on the microwave and start the digestion process according to the following program (Table 1):
5. After digestion, open the vessels in a fume cupboard once they have reached room temperature (*see* **Note 10**).
6. Transfer the samples to a centrifuge tube and fill to the appropriate volume with ultrapure water.
7. If necessary, dilute samples with acidified water 2% (v/v) according to the expected concentration of the elements to be analyzed.
8. All prepared solutions (samples, standards, blanks, etc.) must contain a similar concentration of HNO_3, ranging from 1% to 5% HNO_3 (v/v).
9. If necessary, digested samples should be stored in a cold room (5 ± 3 °C) in PTFE vessels, PFA containers, or PP tubes until analysis.

To troubleshoot any problems that may occur during the digestion process, please refer to **Notes 11–13**.

3.4 ICP-MS Analysis

1. Monitoring room temperature and humidity

   The room temperature should be between 21 ± 3 °C, and it shouldn't vary more than 2 °C per hour.
2. Equipment verification
   - After turning on the equipment, wait at least 30 min for the plasma to stabilize.
   - The verification of the ICP-MS is performed with an optimization solution of 10 μg L^{-1} in 2% HNO_3.

- The following parameters are monitored, and a performance report is produced:
  - Peak width: 0.65–0.85.
  - Peak error: ± 0.05.
  - Sensitivity: A low mass (^{7}Li), an intermediate mass (^{115}In), and a high mass (^{238} U) are monitored to cover the full range of atomic masses analyzed. The minimum signal intensities, measured as counts per second (cps), required are:
    - ^{7}Li: cps > 45 000
    - ^{115}In: cps > 400 000
    - ^{238}U: cps > 800 000
  - Stability: The response variation of at least five consecutive determinations should have a relative standard deviation (%RSD) < 2%.
  - Doubly charged ions: $^{138}Ba^{++}/^{138}Ba < 0.02$.
  - Oxides: $^{156}CeO/^{140}Ce < 0.02$.
- If any criteria of the performance report are not achieved, a mass calibration, autotune, or manual tune should be performed depending on the situation (*see* **Note 14**).
- A mass calibration using a 10 μg L^{-1} multi-element solution should be performed whenever necessary or every 6 months.

3. Interferences

To address the problem of interferences in ICP-MS methodology, two methods may be applied:

- Collision Cell Technology (CCT): This method removes spectral interferences by a collision or reaction gas, allowing lower detection and quantification limits to be achieved. Due to sensitivity requirements, CCT does not apply to all elements; therefore, the choice of strategy may vary depending on the specific element.
- Correction equations: Manufacturer-defined equations are available; however, they may not always be optimal, so an interference study must be conducted. To confirm interference with a specific element and mitigate its effects, solutions with increasing concentrations of potential interferents should be prepared, and the response at the target mass should be measured. If a trend is identified, it must be corrected using an appropriate equation. For validating the correction equations, interferents are spiked into a sample with a known concentration of the element. If the measured concentration remains unchanged across varying interferent

levels, the correction equation is considered appropriate. Whenever a new matrix is analyzed, the correction equations must be studied and introduced in the software if necessary.

4. Calibration curve
   - Calibration should be performed daily, and at least three standards and a blank should be included on each calibration curve.
   - The working ranges are element specific, namely As 0.25–2.5 μg L^{-1}, Cd 0.25–2.5 μg L^{-1}, Pb 0.5–5.0 μg L^{-1} (*see* **Note 15**).
   - The calibration curves are fitted to the experimental points using the equipment's software, which employs the least squares mathematical model.
5. Sample analysis

   The samples, as well as the standards, are placed in 15 or 50 mL disposable tubes in an autosampler.
6. Washing solution

   2% HNO_3, automatically added by the autosampler after each analysis
7. Internal Quality Control
   - Acceptance criteria:
     - HNO_3 distilled in the laboratory: To check for contamination, before the calibration curve, two instrumental blanks should be analyzed, one prepared with HNO_3 distilled in the laboratory (HNO_{3lab}) and the other prepared with commercial 65% ultrapure HNO_3 (HNO_{3com}). HNO_{3lab} is not contaminated if $cpsHNO_{3lab} \leq 1{,}5 \times cpsHNO_{3com}$.
     - Calibration curve: Calibration curves are accepted when the linear correlation coefficient is greater than or equal to 0.9995. The curve points are accepted when they fulfill the $x_{ref} \pm 0.1x_{ref}$ criteria.
   - Internal controls:
     (i) An instrumental blank, followed by internal quality controls, should be analyzed before and after each sample sequence or after every 18 samples.

     (ii) For the internal controls, the accepted values are $x_{ref} \pm 0.2x_{ref}$ for the lowest concentration standard and $x_{ref} \pm 0.1x_{ref}$ for all other standards.

     (iii) A new calibration curve should be analyzed if the acceptance criteria are not met.
     - Internal standard: Recoveries between 80% and 120% are acceptable for the internal standard.

- Measurement of samples and standards: Each tube is measured between 2 and 5 times, and the average value is accepted if the %RSD remains within 10%. If the %RSD exceeds 10%, readings may be discarded, ensuring at least 2 measurements are retained.

- Reference Materials: For reference materials, the accepted values are:
  (i) $x_{ref} \pm$ reference material uncertainty,
  (ii) $x_{ref} \pm$ method uncertainty, whenever the reference material uncertainty is lower than the method uncertainty, for each matrix and element.
    - Replicates: For replicates, the acceptance criterion is expressed as the %RSD, which must not exceed 10%.
    - Fortified samples: Recoveries between 80 and 120% are acceptable for fortified samples.
    - Results: Results should only be reported if they are within the interpolation range of the calibration curve, with extrapolation allowed at the extremes for values up to 10% of the calibration range.
    - Digestion blank: If the result of the digestion blank exceeds the limit of quantification (LOQ) of the method (which corresponds to ten times the standard deviation of the blanks), the blank value should be subtracted from the sample result. If the result of the digestion blank is higher than the first point of the calibration curve, the assay should be repeated.
    - Instrumental blank: If the result of the instrumental blank is higher than the first point of the calibration curve, repeat the measurement of the blank with a new calibration curve. If necessary, retest with ultrapure water from a different source. This criterion should be applied to all instrumental blanks analyzed during the assay.

4 Notes

1. Glassware should not be used to determine trace elements by acid digestion using ICP-MS.
2. The internal standard in ICP-MS is used to monitor the performance of the equipment and correcting variations in sample introduction and ionization. Additionally, it minimizes matrix effects, ensuring greater precision and accuracy in analyses.

Table 2
Blank digestion program for 12 vessels

Step	Power (w)	Temperature (°C)	Time (min)
1	1200	180	5
2	1200	180	15
3	1200	210	5
4	1200	210	10

3. The internal standard can be added individually to each tube before analysis or online using a "T" system.
4. Whenever a new matrix is analyzed, it is necessary to ensure that the concentration of the internal standard in the matrix is negligible. Prepare a 0.25 μg L^{-1} solution with the internal standards. Compare the cps of these elements in the standard solution to the cps of the matrix after digestion. The presence of the internal standard in the matrix is considered negligible if the cps of the matrix is less than 300 cps or is at least 1000 times less than the cps of the 0.25 μg L^{-1} standard.
5. Internal quality control standards must be independent of those used for the calibration curve. For this purpose, commercial solutions should be from different manufacturers or, if from the same manufacturer, from different batches.
6. The program for blank digestion is given in Table 2.

   Please consider the number of vessels to be cleaned and the equipment's specifications. For example, this program is designed for 12 vessels.
7. All vessels must contain the same type and amount of sample, except for the blank. Be sure to set the minimum and maximum volumes in accordance with the manufacturers' specifications, if applicable to the microwave model.
8. After starting the microwave digestion, verify that the monitoring probe rotates freely.
9. If an exothermic reaction is observed during the addition of reagents to the sample, do not immediately close the vessel. Instead, leave it under the fume cupboard and wait until the reaction is finished.
10. If necessary, the vessels can be cooled on ice.
11. Incomplete digestion (visible residues after the cycle) possible cause: Insufficient acid volume or inappropriate temperature. Solution: Increase digestion time, adjust temperature, optimize the reagent composition, or weigh less samples.

12. Vessels breaking or losing seal: Possible causes include excessive pressure due to the use of excessive acid reagent or a highly concentrated sample, improper assembly, or vessel degradation. Solution: Adjust the sample-to-reagent ratio, review digestion steps, check the seal, and avoid exceeding pressure limits, and replace vessels or other consumables as needed.
13. Excessive foaming inside the vessels: Possible cause: High-fat or high-protein samples. Solution: Reduce the sample amount or leave the sample overnight in HNO_3 inside the vessel before adding the remaining reagents and performing the mineralization.
14. If any of the criteria in the performance report are not met, the following corrective actions should be taken depending on the identified issue:
    (i) Mass calibration: This should be performed when issues with peak errors or peak widths are detected. It is also necessary if there are changes in the standard resolution and/or high-resolution parameters.
    (ii) Autotune: Recommended when the instrument response is less sensitive or more unstable than expected. It should also be performed when no cps are detected, as it can automatically optimize the parameters.
    (iii) Manual tune: An alternative for autotune, allowing for specific adjustments.
15. The first point of the calibration curve was defined as being higher than the LOQ of the validation method, which corresponds to ten times the standard deviation of the blanks.

References

1. Cubadda F (2007) Inductively coupled plasma mass spectrometry. In: Food toxicants analysis, pp 697–751. https://doi.org/10.1016/B978-044452843-8/50020-1
2. Agilent Technologies (2006) Inductively coupled plasma mass spectrometry table of contents—ICP-MS Primer. Agilent Technologies, p 84
3. European Standard (2009) EN 15763:2009 – foodstuffs – determination of trace elements – determination of arsenic, cadmium, mercury and lead in foodstuffs by inductively coupled plasma mass spectrometry (ICP-MS) after pressure digestion. European Commission Standard
4. European Standard (2014) En 13805 foodstuffs – determination of trace elements – pressure digestion. European Commission Standard
5. Cubadda F (2004) Inductively coupled plasma-mass spectrometry for the determination of elements and elemental species in food: a review. J AOAC Int 87(1):173–204. https://doi.org/10.1093/jaoac/87.1.173
6. Wilschefski SC, Baxter MR (2019) Inductively coupled plasma mass spectrometry: introduction to analytical aspects. Clin Biochem Rev 40(3): 115–133. https://doi.org/10.33176/AACB-19-00024
7. ATSDR, "Substance Priority List | ATSDR," Agency for Toxic Substances and Disease Registry

8. European Union (2023) Commission Regulation (EU) 2023/465 of 3 March 2023 amending Regulation (EC) No 1881/2006 as regards maximum levels of arsenic in certain foods. Off J Eur Union, L68 7(10):51–53

9. European Union, "COMMISSION REGULATION (EU) 2023/915 of 25 April 2023 on maximum levels for certain contaminants in food and repealing Regulation (EC) no 1881/ 2006," The Official Journal of the European Union

Chapter 5

Ultra-Performance Liquid Chromatography-Tandem Mass Spectrometry (UPLC-MS/MS) for the Determination of Acrylamide in Grains and Grain-Based Products

Inês Delgado, Susana Jesus, Sandra Copeto, and Inês Coelho

Abstract

In recent years, the presence of acrylamide (AA) in food has been a cause of concern, since it has been classified by the IARC as probably carcinogenic to humans (Group 2A). Within the EU, benchmark values were established in 2017, and in 2019, the importance of monitoring this contaminant was further reinforced by a recommendation from the European Commission. The new legislation was the result of a recognition that data on AA levels were still insufficient—not only for foods covered by the 2017 regulation but also for other food categories not covered by the regulation that may nevertheless contain significant levels of AA and/or contribute significantly to overall dietary exposure.

This chapter presents a sensitive, precise, accurate, and straightforward in-house analytical protocol for the determination of AA in cereals and cereal-based products using UPLC–MS/MS. The analytical workflow encompasses critical pre-treatment steps, including sample defatting, aqueous extraction, SPE for sample cleanup. Another critical step is the AA quantification by UPLC-MS/MS, where the extract is then analyzed using a triple quadrupole mass spectrometer operating in ESI^+, with AA-d_3 as an IS. The analyte and IS are separated on an HSS column by gradient elution using water and methanol as the mobile phase. The method achieves a limit of quantification of 19 μg/kg, aligning with regulatory monitoring requirements for AA. Additionally, the main analytical challenges and critical control points throughout the procedure are discussed, with particular emphasis on sample preparation and instrumental parameters that influence quantification accuracy.

Keywords Acrylamide, Analytical methods, Food analysis, Cereals, Process contaminants, Solid-phase extraction

1 Introduction

Acrylamide (AA) is a processing contaminant formed in carbohydrate-rich foods during high-temperature cooking, via the Maillard reaction between asparagine and reducing sugars. It is found in many foods consumed daily around the world, most commonly in French fries, potato chips, bread, cookies, and coffee [1, 2]. The International Agency for Research on Cancer (IARC) has classified

Paula Alvito and Helena Oliveira (eds.), *Food Contaminants Toxicity*, Methods and Protocols in Food Science,
https://doi.org/10.1007/978-1-0716-5096-7_5,

AA as a probable human carcinogen, requiring strict monitoring in food products [3]. In 2015, the European Food Safety Authority (EFSA) published a scientific opinion confirming the carcinogenic properties of this compound [2]. Based on this, in 2017, the European Union (EU) established regulatory measures that set benchmark levels for AA in certain food groups and provided mitigation strategies for the industry to apply when the concentration of this chemical exceeds the benchmark limits [4]. More recently, in 2019, the EU reinforced the importance of monitoring AA levels in specific food groups, such as potato products, bakery products, and cereal products, among others [5].

AA analysis in food is usually performed using liquid chromatography or gas chromatography coupled with mass spectrometry for detection and quantification. Among these, liquid chromatography coupled with tandem mass spectrometry (LC-MS/MS) in multiple reaction monitoring (MRM) mode is widely accepted for AA analysis due to its sensitivity and reliability [6]. However, quantification in complex matrices, such as processed foods, remains challenging, making sample preparation processes crucial for accurate analysis.

The present method developed is an in-house approach based on EN 16618, ensuring accurate quantification of AA in cereal-based matrices using ultra-performance LC-MS/MS (UPLC–MS/MS) with an internal standard (IS) approach for improved accuracy and precision [7]. The protocol includes detailed extraction steps, instrument settings, calibration criteria, and quality assurance measures to ensure compliance with regulatory standards, such as EU Regulation 2017/2158 [4]. The identification and quantification of AA from food matrices involve four main steps: (i) pre-cleaning of the sample, (ii) extraction, (iii) purification, and (iv) identification and quantification. The pre-cleaning step aims to remove food components that interfere with the detection of this contaminant. In this protocol, n-hexane is used to defat the sample. Then, for extraction, water is used, since AA is hydrophilic, and an isotopically labeled AA-d_3 standard is added as IS. In the purification step, solid-phase extraction (SPE) columns are used, with the purpose of retaining as many matrix components as possible while allowing AA to pass through, effectively acting as a chemical filter. The selected SPE column is an ion-exchange column; however, since AA is not retained by the column, the extract passes through and is collected.

Finally, in the last step, AA is identified and quantified by UPLC-MS/MS. For this purpose, a UPLC HSS C18 column is used with gradient elution, using water and methanol as the mobile phase. Detection of AA by UPLC-MS/MS is achieved using the MRM mode with specific transitions. Generally, the most intense transition is used to measure AA levels in food (quantifier), while one or two additional transitions are monitored to confirm signal identity (qualifiers). In this protocol, the transitions used for AA determination are m/z 72 > 55 as the quantifier and m/z 72 > 72, m/z 72 > 44, or m/z 72 > 27 as possible qualifiers. In the case of AA-d_3, the quantifier transition is m/z 75 > 58.

This chapter describes the steps for the determination of AA in cereals and cereal products by UPLC-MS/MS following a validated in-house protocol in accordance with regulatory requirements.

2 Materials

2.1 Materials

1. Centrifuge tubes, of 15 or 50 mL, made of PP.
2. Glass bottles of various volumes.
3. Micropipettes and corresponding tips (*see* **Note 1**).
4. Needles.
5. SPE-SAX columns, or equivalent.
6. Syringes (5 or 10 mL).
7. Waters 12 × 32 mm glass screw neck vial, cap, pre-slit PTFE/ silicone septa, maximum recovery, or equivalent.
8. Waters ACQUITY UPLC HSS C18 column, 1.8 μm, 2.1 × 100 mm, or equivalent.

2.2 Equipment

1. Analytical balance with a resolution of 0.00001 g.
2. Freezer (≤ -18 °C).
3. Knife mill or homogenizer.
4. Mass spectrometer.
5. Refrigerated chamber (5 ± 3 °C).
6. Rotary shaker.
7. Ultra-performance liquid chromatography (UPLC).
8. Vacuum manifold for solid-phase extraction (SPE).
9. Vortex mixer.
10. Water purification system.

2.3 Data Processing

1. MassLynx v4.2 (Waters, Milford, MA, USA).

2.4 Reagents

1. Ultrapure water grade 1(capture from a Milli-Q water purification system).
2. n-Hexane (Merck, $\geq$ 97% suitable for HPLC).
3. Methanol (Merck, LC-MS grade, purity $\geq$99.9%).
4. Formic acid (Group Carlo Erba Reagent, LC-MS-grade, $\geq$ 99%).
5. AA standard (VWR, acrylamide certified standard).
6. AA-d_3 Standard solution (Merck, ~500 mg/L in acetonitrile, analytical standard, for food analysis).

2.5 LC Mobile Phase and Wash Solution Preparation (See Note 2)

1. Mobile phase A (100% ultrapure water grade 1). Replace daily.
2. Mobile phase B (100% methanol).
3. Strong wash (90:10 methanol:water): Add 50 mL of ultrapure water grade 1 and make it to 500 mL with methanol. Mix thoroughly.
4. Weak wash (90:10 water:methanol): Add 50 mL of methanol and make it to 500 mL with ultrapure water grade 1. Mix thoroughly.
5. Seal wash (50:50 water:methanol): Add 100 mL of water and 100 mL of methanol. Mix thoroughly.

2.6 Standard and Internal Standard (See Notes 2 and 3)

1. AA standard solutions: weigh daily 5 mg of AA standard commercial, using a disposable spatula, and make it to 50 mL with ultrapure water grade 1 to obtain S1 solution with a concentration of 100 000 μg/L. See Table 1 for details on the preparation of the following AA standard solutions.
2. AA-d_3 Standard solutions: From the commercial IS solution, stored at 4 °C with a concentration of 500 000 μg/L, prepare a daily solution with a concentration of 5000 μg/L, using ultrapure water (Grade 1) for dilution. Prepare the dilutions according to Table 2.

2.7 Calibrators and Quality Controls (See Notes 2 and 3)

1. The commercial AA standard used for the calibration curve and the standard used for quality controls must be independent. To ensure this independence, the standards either should come

Table 1
Preparation of AA standard working solutions for calibration curves

AA standard solution	Initial concentration (μg/L)	Final concentration (μg/L)	Standard volume (mL)	Final volume (mL)
S2	100 000 (S1)	1000	0.1	10
S3	1000 (S2)	100	0.5	5
S4	100 (S3)	10	0.5	5

Table 2
Preparation of IS working solutions used in the calibration curves and extraction of AA

IS solution	Initial concentration (μg/L)	Final concentration (μg/L)	Standard volume (mL)	Final volume (mL)
IS1	500 000 (commercial solution)	5000	0.1	10
IS2	5000 (IS1)	1500	6	20

from different manufacturers, or if from the same manufacturer, they must be from different batches.

2. The AA can be determined in two working ranges: 1.25–12.5 μg/L or 10–100 μg/L, depending on the initial concentration of AA in the food item. The concentration and preparation of each calibrator are described in Table 3. The calibration curve should be prepared daily and include a minimum of 5 points and one blank.
3. Two quality controls are monitored: one corresponds to the first point of the calibration curve and the other to an intermediate point. The concentration of the controls and the method of preparation are the same as described in Table 3. These controls are prepared daily and are measured after the calibration curve and after the samples.
4. Extraction blank: The blank solution contains all reagents used during the extraction procedure, except for the sample, and follows the same protocol as the sample extraction. The IS solution is also added at the same concentration as the samples.
5. Instrumental blank: The instrumental blank corresponds to the solvent used for the calibration curve, meaning 994 μL of ultrapure water (grade 1), 5 μL of methanol, and 1 μL of formic acid.

3 Methods

3.1 Extraction Procedure (See Notes 2 and 3)

1. Weigh approximately 2 g of homogenized and ground sample into a 50 mL centrifuge tube at least in duplicate (*see* **Notes 4–7**).
2. Add 10 mL of n-hexane.
3. Shake on a mechanical shaker for 5 min (*see* **Note 8**).
4. For fortified samples, add the appropriate AA standard concentration (*see* **Note 9**).
5. Add 0.9 mL of IS solution (IS2) to all samples (*see* **Note 10**).
6. Add 29.1 mL of ultrapure water grade 1, except for fortified samples, where the volume of AA standard added should be subtracted.
7. Shake on a mechanical shaker for 5 min.
8. Centrifuge at 10 000 rpm for 10 min at room temperature and transfer 2 mL of the aqueous phase to a centrifuge tube (*see* **Note 11**).

Table 3
Preparation of calibrators for AA UPLC-MS/MS quantification for low and high working ranges

Working range	Calibrators (μg/L)	AA standard solution Volume (μL)	AA standard solution Solution (Table 1)	Ultrapure water (grade 1) (μL)	Methanol (μL)	Formic acid (μL)	AA-d_3 Standard solutions (IS2) (μL)	Final volume (mL)
Low working range (1.25–12.5 μg/L)	1.25	125	S4	839	5	1	30	1
	1.75	175	S4	789				
	2.5	250	S4	714				
	5	500	S4	464				
	7	700	S4	264				
	10	100	S3	864				
	12.5	125	S3	839				
High working range (10–100 μg/L)	10	100	S3	864	5	1	30	1
	12.5	125	S3	839				
	15	150	S3	814				
	25	250	S3	714				
	50	500	S3	464				
	70	70	S2	894				
	100	100	S2	864				

Table 4
UPLC mobile-phase gradient elution condition for AA determination

Time (min)	Flow rate (mL/min)	Mobile phase A (%)	Mobile phase B (%)
0	0.3	99.0	1.0
4.0	0.3	95.0	5.0
4.5	0.3	10.0	90.0
6.0	0.3	10.0	90.0
6.5	0.3	99.0	1.0
7.5	0.3	99.0	1.0

3.2 Cleanup Procedure (See Notes 2 and 3)

1. Fit the SPE column to the vacuum manifold (*see* **Note 12**).
2. Condition the column with 2.5 mL of methanol, followed by 2.5 mL of ultrapure water grade 1, and discard both portions.
3. Pass 1.5 mL of the aqueous extract through the column and collect the eluate into a 15 mL test tube.
4. Transfer the 994 μL of eluate to a suitable autosampler vial and add 5 μL of methanol and 1 μL of formic acid (*see* **Note 13**).
5. Perform analysis in UPLC-MS/MS.

3.3 Instrumental Operating Conditions

1. Use the UPLC column Waters ACQUITY UPLC HSS C18 (1.8 μm, 2.1 × 100 mm) with the gradient mobile phase as indicated in Table 4 (*see* **Note 14**). The column should be maintained at 30 °C, while the sample manager is kept at 8 °C. The injection volume is 10 μL.
2. Use electrospray ionization (ESI), selected MRM, and unit resolution. Detect AA and the IS with the transitions m/z 72 > 55 and 75 > 58 for quantitative purposes, and m/z 72 > 55, 72 > 54, and 72 > 44 for confirmation of the identity of AA (*see* **Note 15**). The mass spectrometer operating conditions are shown in Table 5 (*see* **Note 16**).

3.4 Data Analysis

1. Data is analyzed using the TargetLynx software (Waters Corporation).
2. With each run, at least a 5-point standard calibration curve is generated for analyte quantification by linear regression of analyte/IS peak area ratio using a 1/x weighting.
3. The limit of detection is 0.4 μg/L and the limit of quantification is 1.25 μg/L.
4. The results are evaluated according to the acceptance criteria described in Table 6.

Table 5
ESI source and MRM acquisition parameters for AA determination

Parameters	AA Quantifier transition 72 > 55	AA Qualifier transition 72 > 72	AA Qualifier transition 72 > 44	AA Qualifier transition 72 > 27	$AA-d_3$ Quantifier transition 75 > 58
Ionization mode	ESI^+	ESI^+	ESI^+	ESI^+	ESI^+
Collision voltage	24	24	24	24	23
Collision energy	10	2	15	15	12
Source temperature	120 °C				
Cone gas flow	30 L/h				
Desolvation temperature	350 °C				
Desolvation gas flow	800 L/h				
Capillary	2.5 kV				

Table 6
Acceptance criteria of the quality control scheme implemented for AA determination (*see* Note 17)

Parameter	Criteria
Retention time	1.30–1.60 min
Calibration curve	$R^2 \geq 0.995$ $x_{ref} \pm 0.10\, x_{ref}$
Quality controls	$x_{ref} \pm 0.10\, x_{ref}$ $x_{ref} \pm 0.15\, x_{ref}$ for lower calibrator
Signal/noise ratio	S/N of quantifier ≥10 S/N of qualifier ≥3 (*see* **Note 18**)
Ion ratio	Calibration curve: ±20% (*see* **Note 19**) Samples and calibrators: Qualifier area/quantifier area should not differ more than ±20% of the following: From the average obtained through longitudinal monitoring of the ratio of the peak areas of the points on the calibration curve or The average of the ratios of the peak areas of the points on the daily calibration curve or The ratio of the calibration standard closest to the sample result on that day
Fortification	75–110%
Duplicates	The evaluation of duplicates is performed in accordance with ISO 5725-6 —Accuracy (trueness and precision) of measurement methods and results.

4 Notes

1. Micropipettes must be checked before any AA analysis or extraction.
2. All the procedures shall be performed in a fume cupboard with adequate airflow.
3. Always keep solutions, standards, and samples protected from light.
4. The sample must be frozen (at - 20 °C) whenever analysis cannot be performed on the same day. On the day of extraction, samples should be thawed and ground using a homogenizer.
5. Ensure that the sample is well crushed and homogeneous, if necessary, you can crush the sample again.
6. The grinder cup must be washed with ultrapure water and ethanol between samples.
7. The evaluation of duplicates is performed in accordance with ISO 5725-6 —Accuracy (trueness and precision) of measurement methods and results. However, this is an internal laboratory criterion, and alternative criteria may be applied.
8. Ensure that the whole sample is in contact with the extractant.
9. For fortified samples, the volume of AA standard to be added depends on the matrix type and its expected AA levels. Usually, for cereals, a spike of 10 μg/L is used, by adding 0.3 mL of S2 (1000 μg/L).
10. It is essential that the IS concentration is the same in both the calibration curve and the samples. In this protocol, the IS concentration is 45 μg/L. Therefore, 0.9 mL of IS2 is added to the samples, considering a final volume of 30 mL.
11. Use a syringe to transfer the aqueous extract. Avoid transferring parts of the fat layer (n-hexane) that will be formed and found on the top, depending on the fat content of the sample.
12. For SPE cleanup, adjust the flow of the SPE columns to allow the liquid to elute drop-wise.
13. Methanol and formic acid are added to the final extract prior to injection to enhance the peak.
14. The response of the UPLC-MS/MS may vary from day to day or over time. Also, the UPLC column may deteriorate after multiple or single use, depending on the number of injections and type of samples analyzed. Therefore, the system should be checked before each series of analyses. Equilibrate the column with the first composition of the gradient of the mobile phase and mass spectrometer for, e.g., 1 h.

15. Depending on the type of UPLC-MS/MS instrument used, other transitions may also be available for peak qualification.
16. Parameters may need to be reoptimized after instrument maintenance.
17. Except for the fortification criteria, all the criteria applied are laboratory specific and have been established based on the particular quality control. However, alternative criteria may be used.
18. As a qualifier transition, only one should be selected each day.
19. The coefficient of variation of the mean of the ionic ratios between the quantifying and the qualifying transitions of the calibration line standards must not exceed 20%.

References

1. Adimas MA, Abera BD, Adimas ZT, et al (2024) Traditional food processing and Acrylamide formation: a review https://doi.org/10.1016/e30258
2. European Food Safety Authority (2015) Scientific opinion on acrylamide in food. EFSA J 13
3. International Agency for Research on Cancer (1994) Agents Classified by the IARC Monographs – Acrylamide
4. European Commission (2017) Commission regulation (EU) 2017/2158: establishing mitigation measures and benchmark levels for the reduction of the presence of acrylamide in food. Off J Eur Union L 304:24–44
5. European Commission (2019) Commission recommendation (UE) 2019/1888 on the monitoring of the presence of acrylamide in certain foods. Off J Eur Union:31–33
6. Kumari A, Bhattacharya B, Agarwal T et al (2023) Method development and validation for acrylamide in potato cutlet by UHPLC-MS/MS. Food Control 151:109817
7. European committee for standardization (2015) EN 16618: food analysis – determination of acrylamide in food by liquid chromatography tandem mass spectrometry (LC-ESI-MS/MS) analyse, Brussels

Chapter 6

Determination of Nitrate and Nitrite in Beet Chips Using UV-Vis Spectrophotometry

David Silva da Costa, Ana Beatriz Porto Fernandes da Silva, Gabriel de Souza Macedo, Luiza Saraiva Bebiano, Camila Akemi Akiyama, Patricia Aparecida de Campos Braga, Juliana Azevedo Lima Pallone, and Adriana Pavesi Arisseto Bragotto

Abstract

Beetroot is a rich source of inorganic nitrate, which, when converted in the body, can provide benefits such as improved physical performance and reduced blood pressure. However, excessive consumption of nitrate may pose health risks due to its conversion to nitrite, which is associated with the formation of carcinogenic compounds and methemoglobinemia. Considering the growing interest in more natural and healthy foods that bring benefits to the body, it is important to evaluate the safety aspects of these products in order to avoid health risks due to the presence of potentially toxic chemical substances. In this sense, this chapter describes an analytical method to assess the presence of nitrate and nitrite in beet chips, in which nitrate is converted into nitrite using spongy cadmium in an alkaline medium and the resulting nitrite is quantified by spectrophotometry at 540 nm after reacting with sulfanilamide and alpha-naphthylamine, forming a colored complex. This research contributes to food safety and nutritional recommendations by providing a useful tool for monitoring nitrate and nitrite levels in foods.

Key words Root vegetables, Contaminants, Snacks, Nitrosamines, Hazards

1 Introduction

Beetroot (*Beta vulgaris* L.) belongs to the Betoideae subfamily and, despite its origin in Europe and North Africa, is widely cultivated in various regions worldwide. Its agronomic development occurs optimally in acidic, deep, and organic matter-rich soils, under mild or cold climatic conditions, which favor its reproductive phase and result in better product quality [1].

This root is notable as one of the primary dietary sources of inorganic nitrate [1]. The concentration of this compound in beetroot is influenced by environmental factors such as atmospheric humidity, temperature, and exposure to natural or artificial light.

Paula Alvito and Helena Oliveira (eds.), *Food Contaminants Toxicity*, Methods and Protocols in Food Science,
https://doi.org/10.1007/978-1-0716-5096-7_6,

Therefore, quality control strategies in production include proper soil fertilization management, nutrient availability, and herbicide use [2].

Nitrates constitute the most common form of nitrogen found in vegetables, with their absorption being intensified by nitrogen fertilization, a technique aimed at optimizing agricultural productivity since nitrogen is essential for plant growth and development [3]. Additionally, curing salts containing sodium and potassium nitrate and nitrite are widely used in the food industry, particularly in meat products, serving as antimicrobial preservatives [4, 5]. However, the primary source of nitrate ingestion in the population comes from vegetable consumption [5].

In recent years, there has been a significant increase in the search for strategies to enhance physical performance, including the intake of nutritional supplements. Recent scientific evidence indicates that the consumption of vegetables rich in inorganic nitrate (NO3-) results in greater bioavailability of nitric oxide (NO), a molecule that plays a crucial role in the physiological regulation of exercise [1]. Nitric oxide is associated with benefits such as reducing the oxygen cost during exercise and increasing exhaustion time, increasing exhaustion [6], promoting vasodilation, and lowering blood pressure [2]. Due to these effects, beetroot juice has been included in the category of "products with the most scientific evidence for improving exercise performance" by the Australian Institute of Sport (AIS) Sports Supplement Framework [1].

However, nitrate ingestion also raises concerns regarding human health. Although nitrate is considered a stable and non-toxic compound at typical consumption levels, a portion of ingested nitrate is reduced to nitrite in saliva by commensal bacteria in the oral cavity, increasing its concentration by more than tenfold [7]. It is estimated that, in healthy individuals, approximately 5–7% of consumed nitrate is converted into nitrite in saliva [8, 9].

Nitrite, in turn, exhibits high chemical reactivity and can interact with erythrocyte hemoglobin to form methemoglobin, a molecule incapable of transporting oxygen. Elevated concentrations of methemoglobin in the body can result in methemoglobinemia, a clinical condition associated with the "blue baby syndrome," particularly in infants [10]. Additionally, nitrite can react with secondary amines to form N-nitroso compounds, such as nitrosamines and nitrosamides, which have carcinogenic potential. Studies indicate that approximately 90% of the nitrosamines evaluated exhibit carcinogenic activity in experimental models and humans [7].

The risks associated with nitrate and nitrite ingestion have been assessed by the Joint FAO/WHO Expert Committee on Food Additives (JECFA), which established an acceptable daily intake (ADI) of 0–3.7 mg/kg body weight (bw) for nitrates and 0–0.07 mg/kg bw for nitrites [11, 12]. These limits were later confirmed by the European Food Safety Authority (EFSA) in its reassessment of the health impacts of these compounds [12, 13]. These ADI values apply to both nitrates and nitrites intentionally added as food additives and those naturally present in foods.

Given this context, the quantitative analysis of nitrates and nitrites in food is essential, especially considering the growing popularity of beetroot products as a nutritional and health option among physically active individuals. Therefore, the objective of this chapter is to describe the criteria and procedures for the determination of nitrate and nitrite in beetroot chips. Nitrate is reduced to nitrite by spongy cadmium in an alkaline medium and the resulting nitrite is diazotized with sulfanilic acid, followed by coupling with alpha-naphthylamine hydrochloride in an acidic medium, forming a rose-colored product. This compound is then quantified spectrophotometrically at 540 nm [14]. Nitrate is determined indirectly, while nitrite is measured directly through the colorimetric reaction. The method allows precise analysis of these compounds in food and can be considered a useful tool for monitoring purposes, enabling a better understanding of the potential benefits and risks associated with the consumption of this product, contributing to food safety and the formulation of more precise dietary recommendations.

2 Materials

This section details the materials used in nitrate and nitrite analysis, focusing on equipment, display cases, and utensils essential for the processes (Table 1). Nitrate and nitrite analysis requires careful and specific preparation of reagents and solutions, which are listed in Table 2.

2.1 Common Materials for Nitrate and Nitrite Analysis

1. 5% Sodium Tetraborate Decahydrate Solution ($Na_2B_4O_7{\cdot}10H_2O$)

   Used as a buffer in both analyses, ensuring pH stability during chemical reactions and preventing variations that could affect the results.

2. 15% (w/v) Potassium Ferrocyanide Trihydrate Solution ($K_4[Fe(CN)_6]{\cdot}3H_2O$)—Carrez 1

   Used to precipitate interfering metal ions in the sample, improving the accuracy of nitrate and nitrite determination.

3. 30% (w/v) Zinc Sulfate Heptahydrate Solution ($ZnSO_4{\cdot}7H_2O$) or Zinc Acetate Dihydrate Solution ($(CH_3COO)_2Zn{\cdot}2H_2O$)—Carrez 2

   Plays a crucial role in removing interfering substances by precipitating impurities and aiding in sample clarification, ensuring greater accuracy in analyses.

Table 1
Equipment, utensils, and glassware used in the analysis of nitrite and nitrate

Category	Material	Details
Equipment	Magnetic stirrer	Used for continuous agitation of solutions and suspensions
	Analytical balance	Precision balance for weighing small quantities of samples
	Food processor	In the preparation of cadmium sponge, the food processor is used to break down agglomerates and reduce particle size, facilitating better reaction and handling
	35-mesh sieve	The 35-mesh sieve is used to separate larger particles, ensuring that only particles smaller than the mesh size (approximately 0.5 mm) pass through, which guarantees an appropriate particle size for the subsequent analysis step
	Water bath	Heated bath for incubation at controlled temperatures
	Shaker	To ensure thorough mixing of the sample and separate solids from the liquid, providing a clear filtrate containing only dissolved components for analysis
	Heating plate	Plate used to heat substances in laboratory containers
	UV-vis spectrophotometer (FLUOstar omega, BMG LABTECH, Germany)	Equipment used to measure light absorption of samples at various wavelengths, specifically for nitrate/nitrite analysis at 540 nm
	pH meter	Device for measuring the pH of a solution
Utensils	Spatula	Used to transfer small quantities of solid substances
	Automatic pipette (2–20 μL; 10–100 μL; 20–200 μL; 100–1000 μL; 1000–5000 μL)	Precision equipment for dispensing liquid volumes accurately
	Funnel	Used to transfer liquids into containers with small openings
	Graduated pipette (5 mL)	Used to measure specific liquid volumes with high accuracy
	pH indicator paper	Used to test the pH of a solution through color change
	Volumetric pipettes (5, 10, 15, 100 mL)	Used to measure liquids with precise volumes
Consumables	Falcon tube (50 mL)	Conical tubes commonly used for centrifugation and storage
	Qualitative filter paper	Filter paper used to separate solid particles from liquids in qualitative analysis
Glassware	Volumetric flasks (1000, 100, 25, 10, 5 mL)	Used to prepare standard solutions with precise volumes
	Beaker (50, 1000, 2000 mL)	Used for mixing, heating, and storing liquids, with various capacities
	Erlenmeyer flask (25 mL)	Narrow-necked flask used for mixing and heating liquids

4. 0.5% (w/v) Sulfanilamide Solution ($C_6H_8N_2O_2$)

   Specifically reacts with nitrites in the sample, producing a colored compound that serves as the basis for their quantification.

5. 0.5% (w/v) Alpha-Naphthyl Ethylenediamine Dihydrochloride Solution ($C_{12}H_{16}Cl_2N_2$)

   Facilitates the formation of colored complexes after reacting with sulfanilamide, enhancing coloration and enabling more sensitive nitrite detection.

2.2 Exclusive Materials for Nitrite Analysis

1. Stock Sodium Nitrite Solution ($NaNO_2$), p.a. Prepare a solution by dissolving 500 mg of sodium nitrite in a 500 mL volumetric flask, then fill to the mark with deionized water. For the working solution, 1 mL is equivalent to 10 μg of sodium nitrite (sodium nitrite stock solution).

2.3 Exclusive Materials for Nitrate Analysis

1. Stock Sodium Nitrate Solution ($NaNO_3$), p.a. Prepare a solution by dissolving 0.1 g of dry sodium nitrate in 50 mL of pH 9.6–9.7 buffer solution. Make the volume to 100 mL with water. For the working solution, 1 mL of this solution corresponds to 10 μg of sodium nitrate. The sodium nitrate stock solution is used in recovery assays to add known amounts of nitrate and verify the accuracy of the extraction process.

2.4 Additional Materials for the Analysis

1. pH 9.6–9.7 Buffer Solution: Dilute 20 mL of concentrated hydrochloric acid (HCl) in 500 mL of water. Add 50 mL of concentrated ammonium hydroxide (NH_4OH) and dilute to 1000 mL with water. Check the pH and adjust if necessary.
2. Diluted pH 9.6–9.7 Buffer Solution (1 + 9): Take 100 mL of the pH 9.6–9.7 buffer solution and dilute to 1000 mL with water.
3. Spongy Cadmium:
   - Prepare a 1000 mL solution of cadmium sulfate (CdSO4.8-H_2O) at 20% concentration and transfer it to a 2000 mL beaker.
   - Add 8–10 metallic zinc bars, ensuring they are fully submerged in the solution. As cadmium deposits on the bars, remove it using a porcelain or plastic spatula and transfer it to a beaker containing enough water to cover the cadmium.
   - Transfer all contents from the beaker into a processing cup (glass or plastic) and use a food processor to break down the agglomerated cadmium, reducing particle size and achieving a finer consistency.
   - Then, filter the mixture through a 35-mesh sieve into another beaker (the 35-mesh designation refers to the number of openings per inch, indicating the particle size

Table 2
Materials and reagents for nitrate and nitrite analysis

Reagent	Preparation/use details
Sodium tetraborate decahydrate ($Na_2B_4O_7{\cdot}10H_2O$)	5% solution, used as a buffer to stabilize pH in analyses
Potassium ferrocyanide trihydrate ($K_4[Fe(CN)_6]{\cdot}3H_2O$)	15% (w/v) solution, it forms colored complexes with nitrate and nitrite
Zinc sulfate heptahydrate ($ZnSO_4{\cdot}7H_2O$) or zinc acetate dihydrate ($(CH_3COO)_2Zn{\cdot}2H_2O$)	30% (w/v) solution, it removes impurities by precipitating interfering substances
Sulfanilamide ($C_6H_8N_2O_2$)	0.5% (w/v) solution, it reacts with nitrite forming colored compound
Alpha-naphthyl ethylenediamine Dihydrochloride ($C_{12}H_{16}Cl_2N_2$)	0.5% (w/v) solution, it reacts with sulfanilamide to form colored complexes
Deionized water	Used to prepare nitrate and nitrite solutions, buffer solutions, and for washing the cadmium
Sodium nitrite ($NaNO_2$)	Stock solution: 500 mg dissolved in 500 mL (1 mL = 10 μg $NaNO_2$)
Sodium nitrate ($NaNO_3$)	Stock solution: 0.1 g dissolved in 50 mL buffer pH 9.6–9.7, diluted to 100 mL (1 mL = 10 μg $NaNO_3$)
Hydrochloric acid (HCl)	Used to prepare buffers and to treat sponge cadmium
Ammonium hydroxide (NH_4OH)	Used to prepare buffers at pH 9.6–9.7
Metallic zinc	Used in the preparation of sponge cadmium by reacting with cadmium sulfate
Cadmium sulfate ($CdSO_4{\cdot}8H_2O$)	20% solution, used to prepare sponge cadmium
Metallic zinc (solid bars)	Solid zinc bars used to prepare cadmium sponge by reduction with cadmium sulfate
Sodium hydroxide	20% solution, used to adjust the sample pH to alkaline, enabling proper reaction with sulfanilamide. The 20% concentration ensures optimal pH without destabilizing the diazo complex or causing interference

cutoff of the sieve) keeping the cadmium submerged in water. The material retained in the sieve should be returned to the 2000 mL beaker with zinc bars for later reuse. Allow the beaten and sieved cadmium to settle completely.

- Remove the supernatant water and begin processing the cadmium residue (spongy cadmium).
- Cover the cadmium with 2 N hydrochloric acid (HCl) and let it rest for no more than 2 min.

- Decant and remove the supernatant HCl. Wash the cadmium with deionized water until it reaches neutral pH or the pH of the washing water, using pH indicator paper.
- Add 0.1 N hydrochloric acid (HCl) solution to cover the cadmium, allowing it to rest for at least 15 min.
- Decant and remove the supernatant HCl.
- Wash the cadmium with water (MilliQ water) until it reaches neutral pH.
- Decant and remove the supernatant water and add the diluted pH 9.6–9.7 buffer (1 + 9). Let it rest for at least 15 min. After this time, the cadmium is ready for use.
- During the analysis, the cadmium should remain fully submerged in the diluted buffer solution to ensure proper function. After use, it must be carefully washed with water to remove impurities and then kept continuously submerged in water while resting. This procedure prevents oxidation and maintains its effectiveness. When cadmium recovery is needed, the same treatment process described earlier for used cadmium should be followed to prepare it for reuse.
- Unused spongy cadmium should be kept in water. Replace the water with the diluted pH 9.6–9.7 buffer (1 + 9) at least 15 min before use in analyses.

3 Methods

3.1 Procedure for Standard Curve Preparation and Measurement

1. Pipette aliquots of 50, 100, 200, 250, 300, 350, and 400 μL from the 10 μg/mL sodium nitrite stock solution into separate 5 mL volumetric flasks (Fig. 1). The resulting concentrations of the standard solutions will be approximately 0.1, 0.2, 0.4, 0.5, 0.6, 0.7, and 0.8 μg/mL, respectively. Always add 1 mL of deionized water to each volumetric flask before adding the aliquots to prepare the calibration curve.
2. The limit of detection (LOD) and limit of quantification (LOQ) were determined based on visual perception of the analytical response, using successive dilutions of standard solutions until the minimum concentration distinguishable from the blank was identified. For nitrite, the LOD was established at 12.5 μg/g and the LOQ at 25 μg/g. For nitrate, the LOD was 125 μg/g and the LOQ was 250 μg/g, representing the lowest concentrations at which the analyte signal could be confidently detected and quantified. This procedure followed the visual evaluation approach, ensuring clear differentiation from the blank [15].

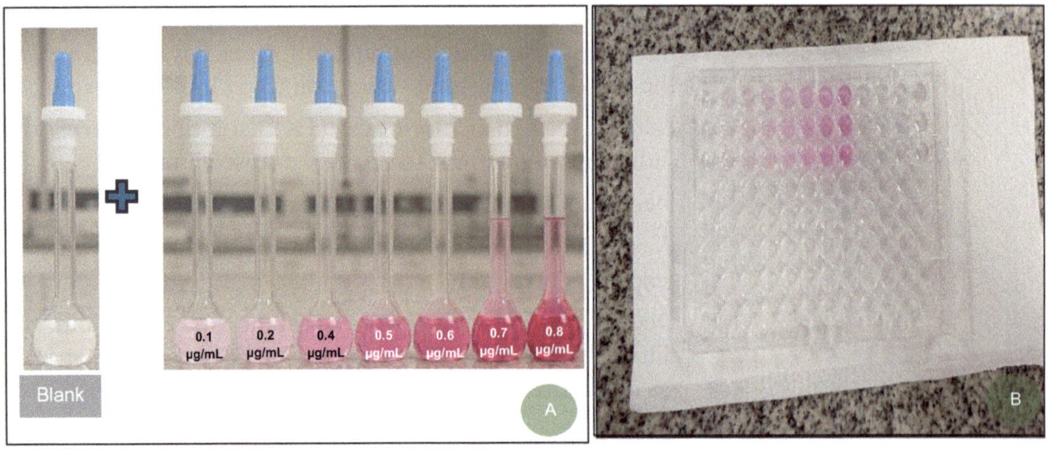

Fig. 1 (**a**) Nitrite calibration curve and (**b**) 96-well plate for reading on the spectrophotometer. (Source: Author)

3. To each flask, add 500 µL of a 0.5% sulfanilamide solution. Wait for 3 min before adding 300 µL of a 0.5% alpha-naphthylethylenediamine chloride solution, shaking after each addition.
4. Fill the volumetric flasks up to the mark with deionized water and mix well. Leave the solutions to stand for 30 min;
5. Measure the absorbance of each solution at 540 nm against a reagent blank.
6. This procedure will generate a linear standard curve with an R^2 value greater than 0.98 [16] (Fig. 2), which is considered acceptable for accurate quantification of nitrite and nitrate concentrations.

3.2 Nitrite and Nitrate Analysis

3.2.1 Common Procedure for Nitrate and Nitrite Determination

1. Weigh 1 g of the homogenized sample in a 50 mL Falcon tube. Homogenize using a magnetic stirrer for 1 min. Add 15 mL of deionized water heated to 60 °C and homogenize again for 1 min. Then, add 1000 µL of a 0.5% sodium tetraborate solution (*see* **Note 1**). Place the mixture in a water bath for 30 min, agitating every 2 min. Using a funnel and spatulas, quantitatively transfer the contents of the Falcon tube to a 25 mL volumetric flask.
2. Add 1000 µL of a 15% potassium ferrocyanide solution (*see* **Note 2**) and 1000 µL of a 30% zinc sulfate (*see* **Note 3**). Agitate by rotation after adding each reagent and complete the volume with heated deionized water (60 °C) (*see* **Note 4**). Filter the solution through qualitative filter paper to a 50 mL beaker (Fig. 3).

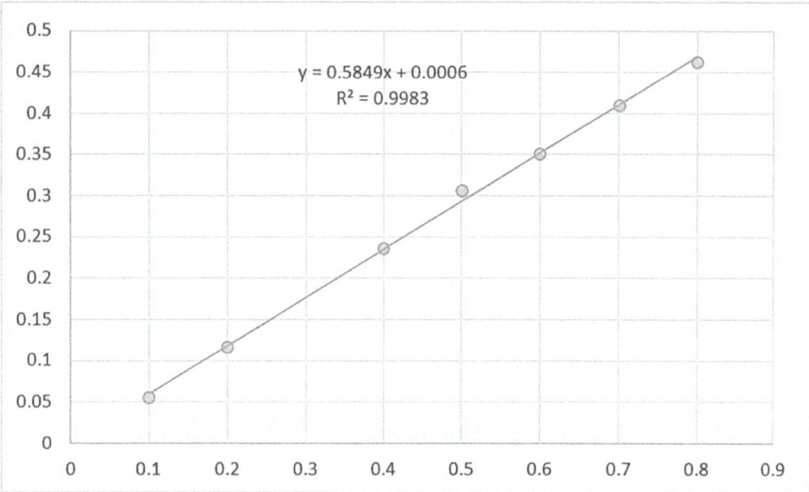

Fig. 2 Calibration curve for nitrite, where the x axis represents the nitrite concentrations, ranging from 0.1 to 0.7 μg/mL, and the y axis shows the average absorbance (triplicates) measured at 540 nm. This calibration curve was used to determine the nitrite concentration in unknown samples by comparing their absorbance values to the established curve. (Source: Author)

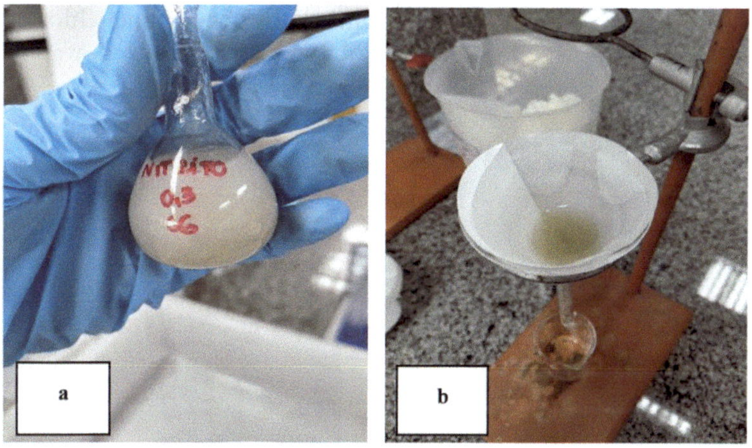

Fig. 3 (**a**) Sample deproteinization step and (**b**) filtration step for obtaining the deproteinized filtrate. (Source: Author)

3.2.2 Quantitative Analysis of Nitrite

1. Transfer 1 mL of the filtrate into a 5 mL volumetric flask. Add 100 μL of 20% sodium hydroxide solution (*see* **Note 5**) and let it react for 10 min.
2. Add 500 μL of 0.5% sulfanilamide solution (*see* **Note 6**) and let it react for 3 min. Then, add 300 μL of 0.5% alpha-naphthylethylenediamine hydrochloride solution (*see* **Note 7**), agitating after each addition.
3. Complete the volume with deionized water and homogenize (*see* **Note 8**). Let it rest for 30 min. After this period, measure

the absorbance at 540 nm at 37 °C using a 96-well plate with a volume of 300 μL. Perform a corresponding blank measurement (*see* **Note 9**).

3.2.3 Quantitative Analysis of Nitrate

1. Transfer 2 mL of the deproteinized filtrate into a 25 mL Erlenmeyer flask. Add 500 μL of the buffer solution with pH between 9.6 and 9.7 and approximately 1 g of spongy cadmium (Fig. 4) to each flask.
2. Place the flask on a shaker at speed 2 for a minimum of 15 min, and then allow it to sediment. Filter the supernatant through qualitative filter paper, collecting it directly in a 10 mL volumetric flask.
3. Wash the Erlenmeyer flask at least 3 times with 2 mL of water, shaking for 2 min after each wash. Let it sediment and filter.
4. Wash the filter paper and complete the volume of the 10 mL volumetric flaskwith water. Pipette 0.5 mL in a 5 mL volumetric flask.
5. Add 500 μL of the buffer solution with pH between 9.6 and 9.7, followed by 500 μL of the 0.5% sulfanilamide solution, and mix. After 3 min, add 300 μL of the 0.5% α-naphthylethylenediamine chloride solution and complete the volume with deionized water. After 30 min, read the

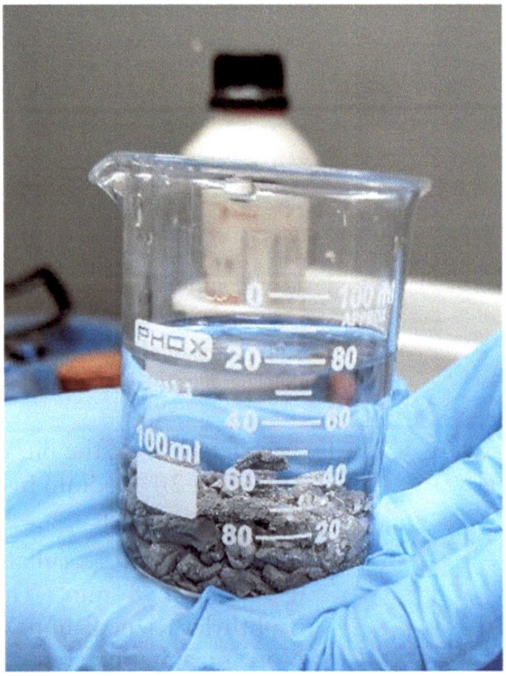

Fig. 4 Preparation of sponge cadmium—before undergoing the acid treatment process and pH 9.6–9.7 adjustment. (Source: Author)

absorbance at 540 nm at 37 °C. The result corresponds to the total nitrite content.

3.3 Calculations for Nitrite and Nitrate Concentration

3.3.1 Nitrite Concentration

1. The nitrite concentration (X) can be calculated by rearranging the equation of the line:

$$X = Y - b/a$$

Where:

- Y is the absorbance measured for the sample
- b is the y-intercept
- a is the slope (absorptivity).

Considering the dilution during sample preparation, the concentration of nitrite in the sample (μg/g) is adjusted based on the dilution factor:

$$X \text{ nitrite } (\mu g/\text{g}) = x \text{ measured } (\mu g/mL) \times 125/1$$

Where:

- x: nitrite concentration in μg/mL
- 125: the number of times a sample is diluted
- 1: sample weight in g

3.3.2 Nitrate Concentration

To calculate the nitrate concentration from the nitrite concentration, the following conversion factor is used:

$$\text{Total nitrite : The total nitrite in } \mu g/\text{g} = \text{x}^{*}\ 1250/1$$

Where:

- x: the total nitrite concentration in μg/mL.
- 1250: the number of times the sample is diluted during nitrate extraction.
- 1: the sample weight in grams.

Nitrate is then calculated from the difference between total nitrite (nitrite plus reduced nitrate) and nitrite:

$$\text{Nitrate concentration } (\mu g/\text{g}) = (\text{total nitrite } (\mu g/\text{g}) - \text{nitrite } (\mu g/\text{g})) \times 1.231$$

Where:

- 1.231 is the conversion factor of nitrites to nitrates [14].

Figure 5 presents a diagram of all the steps involved in the determination of nitrite and nitrate in beet chips.

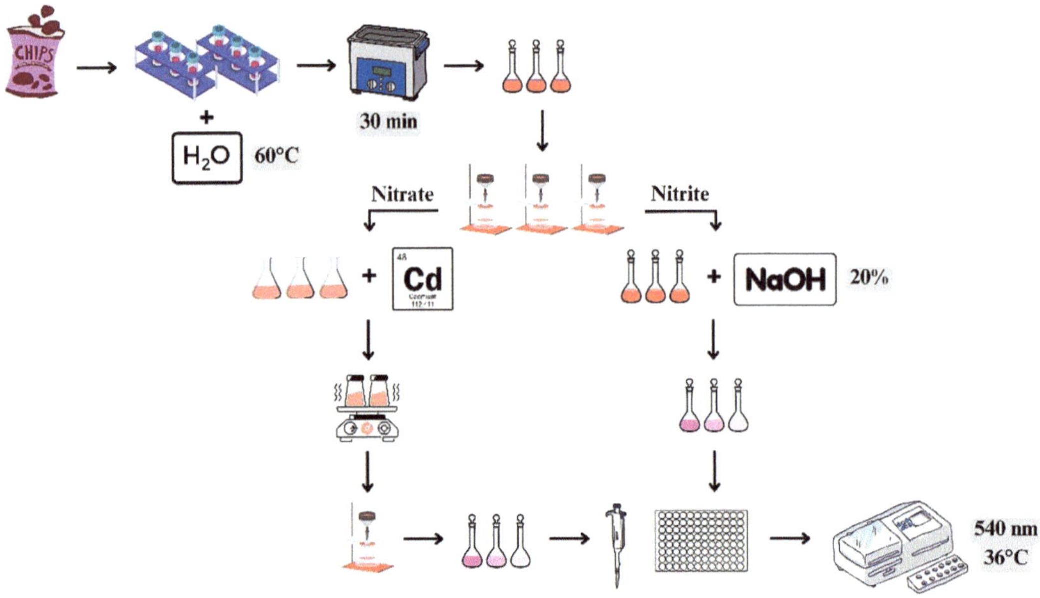

Fig. 5 Summary of all steps in the quantification of nitrate and nitrite. (Source: Author)

4 Notes

1. Sodium Tetraborate (0.5%): Sodium tetraborate is employed to remove interfering substances from the sample, such as proteins, lipids, and insoluble compounds, through complex formation. This facilitates the removal of these materials, ensuring their minimal interference in subsequent analyses. A 0.5% concentration is optimal for efficient complexation without excessive precipitation, which could impede filtration or analytical accuracy. Higher concentrations may lead to unwanted precipitation, while lower concentrations might fail to adequately remove contaminants.
2. Potassium Ferrocyanide 15% (Carrez 1): Potassium ferrocyanide is utilized to stabilize sample components and prevent unwanted precipitation during the analytical process. It selectively binds to heavy metals and other potential interferents, preventing their interference with the analysis of target compounds, such as nitrites. The 15% concentration is sufficient to react with metal ions present in the sample without excessive precipitation or altering the sample's overall composition. Suboptimal concentrations could be ineffective at sequestering metals, while excessive amounts might form solid complexes, complicating analysis.
3. Potassium Sulfate 30% (Carrez 2): Potassium sulfate acts as a precipitant, facilitating the removal of proteins and other water-soluble contaminants, thus enhancing the clarity of the

solution. A 30% concentration is chosen for its efficacy in promoting precipitation without over-precipitating, which could compromise recovery or interfere with further analysis. Concentrations that are too low may not sufficiently precipitate interferents, while higher concentrations may lead to the loss of valuable sample components.

4. Ultrapure Water at 60 °C: The use of ultrapure water heated to 60 °C is essential in promoting the dissolution of reagents and enhancing the extraction of compounds from the sample. The elevated temperature accelerates the reaction kinetics, ensuring efficient and uniform mixing. The use of ultrapure water eliminates the risk of introducing ions or contaminants that could interfere with the analysis, preserving the integrity of the nitrite determination. The controlled temperature also aids in dissolving more resistant substances, preventing incomplete reactions that could lead to inaccuracies.
5. Sodium Hydroxide (NaOH, 20%): Sodium hydroxide is used to adjust the pH of the sample to an alkaline condition, which is necessary for the proper reaction with sulfanilamide. A 20% NaOH solution provides the required pH adjustment without causing excessive alkalinity, which could negatively affect the stability of the diazo complex and introduce interference. Higher concentrations could lead to overly alkaline conditions, disrupting complex formation, while lower concentrations may fail to sufficiently alter the pH for optimal reaction conditions.
6. Sulfanilamide Solution (0.5%): Sulfanilamide plays a critical role in the formation of the diazo complex with nitrite present in the sample. A 0.5% concentration ensures an adequate amount of sulfanilamide to efficiently form the complex with nitrites, providing reliable analytical results. Higher concentrations could lead to unwanted precipitation or interfere with subsequent reactions, while lower concentrations may not provide sufficient reagent to ensure accurate complex formation.
7. Potassium Alpha-Naphthylethylenediamine Chloride Solution (0.5%): Potassium alpha-naphthylethylenediamine chloride is essential for the development of the characteristic color of the diazo-sulfanilamide complex, which is measured at 540 nm. The 0.5% concentration was selected to ensure efficient color formation without excess reagent, which could result in precipitation or inaccurate readings. Higher concentrations could lead to excessive color formation, complicating the measurement, while lower concentrations may not produce a sufficiently strong signal for precise quantification.
8. Nitrite Determination: Nitrites are quantified from the deproteinized filtrate, as per the established nitrite analytical methodology.

9. Nitrite-Free Water: It is crucial to use nitrite-free water throughout the process to prevent any interference with the analytical results.

References

1. Bescos R, Rollason ML, Davies TS, Casas-Agustench P (2023) Nitrate and nitrite content in commercial and homemade beetroot juices and the effect of storage temperature. Food Sci Nutr 11:6376–6383. https://doi.org/10.1002/fsn3.3575
2. Baião D, da Silva DVT, Paschoalin VMF (2020) Beetroot, a remarkable vegetable: its nitrate and phytochemical contents can be adjusted in novel formulations to benefit health and support cardiovascular disease therapies. Antioxidants 9:960. https://doi.org/10.3390/antiox9100960
3. Pôrto MLA et al (2012) Productivity and nitrate accumulation in zucchini fruits as a function of nitrogen fertilization. Bragantia 71:190–195. https://doi.org/10.1590/S0006870520120050000020
4. Bian Z et al (2020) A review of environment effects on nitrate accumulation in leafy vegetables grown in controlled environments. Foods 9:732. https://doi.org/10.3390/foods9060732
5. Hmelak Gorenjak A, Cencič A (2013) Nitrate in vegetables and their impact on human health. A review. Acta Aliment 42(2): 158–172. https://doi.org/10.1556/aalim.42.2013.2.4
6. Dhiman A, Suhag R, Chauhan DS, Thakur D, Chhikara S, Prabhakar PK (2021) Status of beetroot processing and processed products: thermal and emerging technologies intervention. Trends Food Sci Technol 114:443–458. https://doi.org/10.1016/j.tifs.2021.05.042
7. Moreno C, Brayan O, Soto O, González R (2015) Nitrate consumption and its potential beneficial effect on cardiovascular health. Rev Chil Nutr 42:199–205. https://doi.org/10.4067/S0717-75182015000200013
8. Chan TYK (2011) Vegetable-borne nitrate and nitrite and the risk of methemoglobinemia. Toxicol Lett 200:107–108. https://doi.org/10.1016/j.toxlet.2010.11.002
9. Pannala AS, Mani AR, Spencer JPE, Skinner V, Bruckdorfer KR, Moore KP, Rice-Evans CA (2003) The effect of dietary nitrate on salivary, plasma, and urinary nitrate metabolism in humans. Free Radic Biol Med 34:576–584. https://doi.org/10.1016/S0891-5849(02)01353-9
10. Fossen Johnson S (2022) Nitrates and methemoglobinemia. In: Nitrate handbook. Taylor & Francis. https://doi.org/10.1201/9780429326806-20
11. Speijers G, Van den Brandt PA (2003) Nitrite and potential endogenous formation of N-nitroso compounds; safety evaluation of certain food additives, JECFA, WHO food additives series, 50. WHO, pp 49–74
12. European Food Safety Authority (EFSA) (2017a) Re-evaluation of sodium nitrate (E 251) and potassium nitrate (E 252) as food additives. EFSA J 15:4787. https://doi.org/10.2903/j.efsa.2017.4787
13. European Food Safety Authority (EFSA) (2017b) Re-evaluation of potassium nitrite (E 249) and sodium nitrite (E 250) as food additives. EFSA J 15:4786. https://doi.org/10.2903/j.efsa.2017.4786
14. Ministry of Agriculture and Livestock (MAPA) (2022) Official analytical methods for the control of animal-derived products and their ingredients. National Secretariat of Agricultural Defense, National Animal Reference Laboratory. Available at: https://www.gov.br/agricultura/pt-br/assuntos/lfda/legislacao-metodos-da-redelfda/poa/metodos_oficiais_para_analise_de_produtos_de_origem_animal-1aed2022assinado.pdf
15. National Institute of Metrology, Standardization, and Industrial Quality (INMETRO) (2020) Guidance on validation of chemical testing methods. DOQ-CGCRE-08, Revision 09, April 2020
16. Guideline IHT (2005). Validation of analytical procedures: text and methodology Q2 (R1), 1(20), 05

Chapter 7

Heavy Metal Phytotoxicity Evaluation: Method for Antioxidant Profiling in Leaves

Marcelo D. Catarino, Susana M. Cardoso, Sónia Silva, and Maria Celeste Dias

Abstract

Heavy metal environmental contamination severely limits crop growth and productivity due to its high phytotoxicity. One of the primary mechanisms by which heavy metals affect plants is the generation of reactive oxygen species, which can damage biomolecules or function as signaling molecules that activate antioxidant defenses. Antioxidant compounds play a crucial role in mitigating oxidative damage and alleviating the negative effects on plant physiology and performance. This chapter presents a method for extracting phenolic compounds and characterizing them using liquid chromatography to assess heavy metal-induced phytotoxicity. Phenolic profiling is a useful approach to evaluate plant antioxidant capacity under heavy metal stress and to identify compounds with relevant role in stress response.

Key words Phenolic compounds extraction, Ultra-high performance liquid chromatography, mass spectrometry, Reactive oxygen species, Oxidative stress, Toxicity

1 Introduction

Heavy metal pollution is a widespread global issue, disrupting ecosystems and posing severe health risks to humans [1]. In agricultural areas, heavy metal contamination affects the food chain, posing a significant threat to food security [2]. The accumulation of heavy metals in soil and their subsequent uptake by plants cause various physiological impairments, severely hindering plant growth and productivity due to their high phytotoxicity [1–6]. One of the first mechanisms by which heavy metals affect plants performance is through the induction of excessive production of reactive oxygen species (ROS). When ROS levels exceed the tolerance threshold, they can cause oxidative stress leading to cellular damage. In turn, ROS can serve as signaling molecules that activate antioxidant defenses [7]. Plant responses to metal toxicity involve a wide range of secondary metabolites [8]. These secondary metabolites

Paula Alvito and Helena Oliveira (eds.), *Food Contaminants Toxicity*, Methods and Protocols in Food Science,
https://doi.org/10.1007/978-1-0716-5096-7_7,

can directly scavenge ROS due to their ability to donate hydrogen atoms or electrons [8]. Phenolic compounds are a good example of such secondary metabolites that play a crucial role in mitigating heavy metal toxicity, acting as antioxidants, metal chelators, and signaling molecules [9]. Flavonoids are among the most abundant and diverse phenolic compounds found in plant tissues. This group includes compounds like flavones, flavan-3-ols, flavonols, and anthocyanins and are usually responsible for the blue, purple, yellow, orange, and red colors in multiple plant tissues [10]. Phenolic acids such as hydroxycinnamic and hydroxybenzoic acids also stand out as key compounds in plant species, being enrolled in the protection against stresses such as infections, wounding, and UV radiation [11]. Thus, monitoring phenolic abundance and profile can provide valuable insights into the severity of heavy metal contamination and the potential resilience of plants to environmental pollutants [3, 12]. This makes phenolic compounds reliable biomarkers for evaluating phytotoxicity in ecological and environmental studies.

The most preferred and consistent analytical technique for separation, identification, and quantification of phenolic compounds is via high performance liquid chromatography (HPLC) [13]. In general, phenolics are analyzed in HPLC instruments equipped with reverse-phase C18 columns (RP-C18) coupled to a photo diode array detector and/or a mass spectrometer (HPLC-DAD-MS), and using a binary gradient elution system composed of an acidified solution and a polar solvent (commonly formic acid and acetonitrile, respectively) [11]. Advances in separating technologies brought, among others, the UHPLC instruments which are capable of operating in higher pressures and consequently improve the analysis efficiency in much shorter timings [14].

Currently, a wide range of chromatographic methodologies for the analysis of phenolic compounds has been extensively documented in the literature. These methodologies vary in terms of sample preparation techniques, extraction protocols, detection systems, and chromatographic conditions, allowing for the tailored analysis of phenolics in complex matrices. Such versatility in analytical methodologies reflects the diversity of phenolic compounds and the complexity of plant matrices, underscoring the importance of selecting appropriate techniques based on specific research objectives.

Therefore, this work aims to describe a methodology suitable for the extraction and analysis of phenolic compounds in plant leaves to assess heavy metal-induced phytotoxicity. This method seeks to offer reliable and accurate results that can be used to evaluate the impact of heavy metals on plant health and stress responses, ultimately contributing to the development of effective biomarkers for environmental monitoring. In addition, this methodology can also be applied to plants exposed to other environmental stressors, such as drought, heat, salinity, or UV radiation, where antioxidant responses are relevant.

2 Materials

Prepare all material and analytical grade reagents. The description provided for sample preparation and extraction refers to only one sample extraction. Strictly adhere to all waste disposal regulations when discarding waste materials.

2.1 Preparation of samples and extraction of phenolic compounds

Sample preparation: Small mill, precision balance, laboratory spoon, 5 g of dry leaves (dried for 1 week at 40 °C), and a vortex (*see* **Note 1**).

Metabolite Extraction: One Erlenmeyer (250 mL or 100 mL), 110 ml of methanol (purity ≥99.8%), two squares of parafilm (6 cm × 6 cm), 40 cm of aluminum foil, digital orbital shaker, one evaporating flask (ST/NS 29/42) of 250 mL, rotatory evaporator, one volumetric flash of 10 mL, one spatula spoon (21 cm).

Analysis: One syringe of 1 mL, one nylon membrane filter with a pore size of 0.2 μm, one chromatography screw thread vial of 2 mL.

Material for alternative method suitable for limited plant material availability:

Sample preparation and metabolite extraction: 5 mL of methanol (purity ≥99.8%), 100 mg of fresh leaf material, 1 L of liquid nitrogen, pestle and a small mortal, 1.5 mL Eppendorf tubes or small glass tubes, vortex, ultrasonic bath, and centrifuge for Eppendorf tubes.

Analysis: One syringe of 1 mL, one nylon membrane filter with a pore size of 0.2 μm, one chromatography screw thread vial of 2 mL with inserts (volume of 0.15 mL).

2.2 UHPLC-DAD-ESI-MS apparatus and conditions for the analysis of phenolic compounds

Apparatus: Ultimate 3000 UHPLC apparatus, consisting of an autosampler, a binary pump, a column compartment, and an Ultimate 3000 diode array detector, coupled to an ion trap mass spectrometer, with an electrospray ionization source (ESI), for example, a Thermo LTQ XL. Other apparatus with similar specifications can be used.

Chromatographic column: C18 column, for example, a Hypersil Gold, with 100 mm length, 2.1 mm i.d.,1.9 μm particle diameter and end-capped; or other with similar specifications.

Solvents and gases: Acetonitrile for the polar eluent (solvent B) and 0.1% formic acid solution for the acidified eluent (solvent A), nitrogen, and helium. All solvents and gases must be high-purity HPLC-MS grade (>99%). The eluents must be degassed before use and kept in tight sealed bottles at room temperature.

Instrument settings: Column oven temperature should be set to 30 °C, ESI needle voltage to 5.00 kV, the capillary temperature maintained at 275 °C, and nitrogen gas pressure at 520 kPa. For CID-MS/MS experiments, helium collision energy should be set to 25–35 arbitrary units.

2.3 Semi-quantification of compounds

Standard compounds preparation: Pure compounds for calibration curves preparation, methanol (purity ≥99.8%), volumetric flasks, precision balance and laboratory spoons (see Note 2).

Analysis: 1 Syringe of 1 mL, 1 nylon membrane filter with a pore size of 0.2 μm, 1 chromatography screw thread vial of 2 ml.

3 Methods

Carry out all procedures at room temperature (<25 °C) unless otherwise specified.

3.1 Sample Preparation

Grind 5 g of dry plant leaves (*see* **Note 3**) in a small mill. When the leaves are completely crushed into powder, weigh 5 g and add to the Erlenmeyer flask.

Add 50 mL of methanol (*see* **Note 4**) into the Erlenmeyer flask with the leaf powder.

Close the Erlenmeyer flask with parafilm and cover it with aluminum foil to protect from the light.

Alternative method suitable for limited plant material availability (*see* **Note 5**) [11]:

Weight 100 mg of fresh leaf material ground in liquid nitrogen and transfer it to a glass tube or an Eppendorf.

Add 1 mL methanol to the tube containing the ground leaf material.

3.2 Extraction Process

Put the Erlenmeyer in an orbital shaker at 200 rpm for 24 h. After this period, turn off the digital orbital shaker and leave the Erlenmeyer for 30 min in order to get a pellet (leaf tissues). Very carefully, open the Erlenmeyer and pour the upper phase (methanol with the metabolites) into an evaporating flask. After the first extraction cycle, proceed with a second extraction of the pellet by repeating the same procedure.

Evaporate the upper phase (methanol with the metabolites) obtained in the two extraction cycles to dryness at low pressure in a rotatory evaporator. Leave the extract in the evaporating flask to dry for 7 days in a dry and dark protected space.

Put a volumetric flash in a scale. Remove the extract in the evaporating flask with a spatula and weight 1 g of the extract in the volumetric flash. Add 10 mL of methanol. Shake until the complete dissolution of the extract in the methanol. If necessary, homogenize the sample using a vortex.

With a syringe, remove 1 mL of the extract and filter into a chromatography vial. Close the vial with silicone screws (*see* **Note 6**).

Alternative method suitable for limited plant material availability:

Homogenize the sample using a vortex for 30 s.

Sonicate the samples at 40 °C for 30 min.

Centrifugate at 15,000 g for 15 min.

Transfer the supernatant for a clean glass tube suitable for centrifugation or an Eppendorf.

Using a syringe, collect at least 150 μL of the extract and filter it through a 0.2 μm nylon membrane into a chromatography vial with an insert. Close the vial with silicone screws (*see* **Note 7**).

3.3 Analysis and characterization of phenolic compounds

Set the temperature of the automatic sampler compartment at 4 °C and put the vials with samples inside.

Run the analysis using the following conditions:

Injection volume: 8 μL.

Flow rate: 0.2 μL/min.

Gradient program: 5–40% of solvent be from 0–14.72 min; from 40–100% of solvent B from 14.72 to 16.63 min; maintain at 100% from 16.63 to 19.54 min; return to 5% from 19.54–34 min.

Recording of the chromatographic profiles must be done at 280 nm, and UV-vis spectral data should be acquired between 200 and 700 nm.

Operate the MS in negative mode and scan over an m/z range from 100 to 2000.

The resulting chromatograms will show peaks with different intensities expressed in absorbance units (AU) that correspond to the multiple phenolic compounds present in the samples. These can be identified by comparing their retention times, UV spectra, and deprotonated molecular ions with that of standard phenolic compounds, when available, injected under the same conditions, or by comparing the UV spectra, deprotonated molecular ions, and fragmentation patterns with what has been previously reported in the literature. Figure 1 displays an illustrating chromatographic profile of a sample run under the conditions mentioned above, alongside with the spectrophotometric and mass spectrometric data for the major compound.

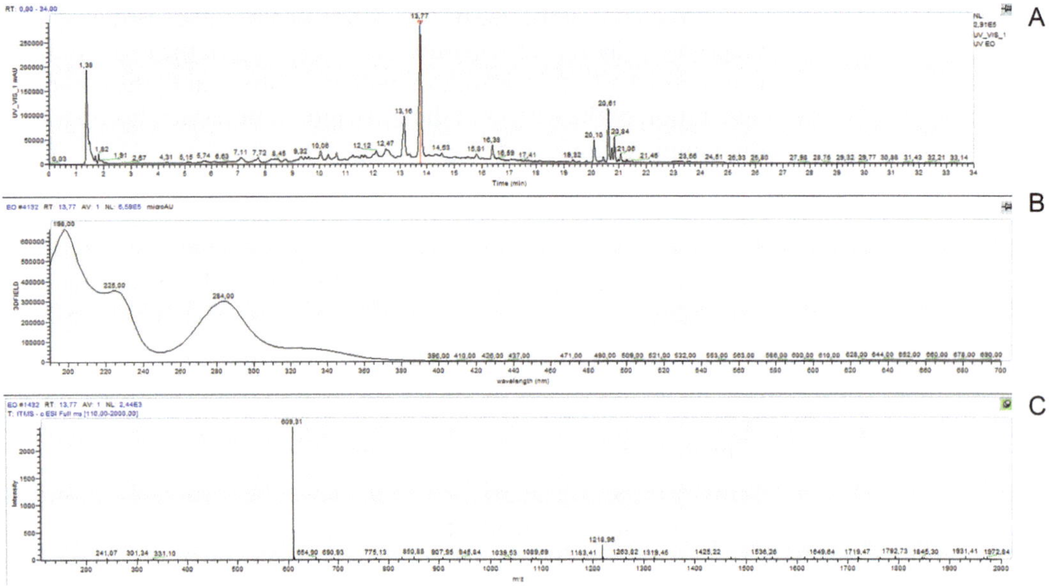

Fig. 1 Polyphenol chromatographic profile of a sample recorded at 280 nm (**a**) accompanied by the UV-vis spectrophotometric information of the major peak detected (**b**) as well as the mass spectroscopic information for that same peak (**c**)

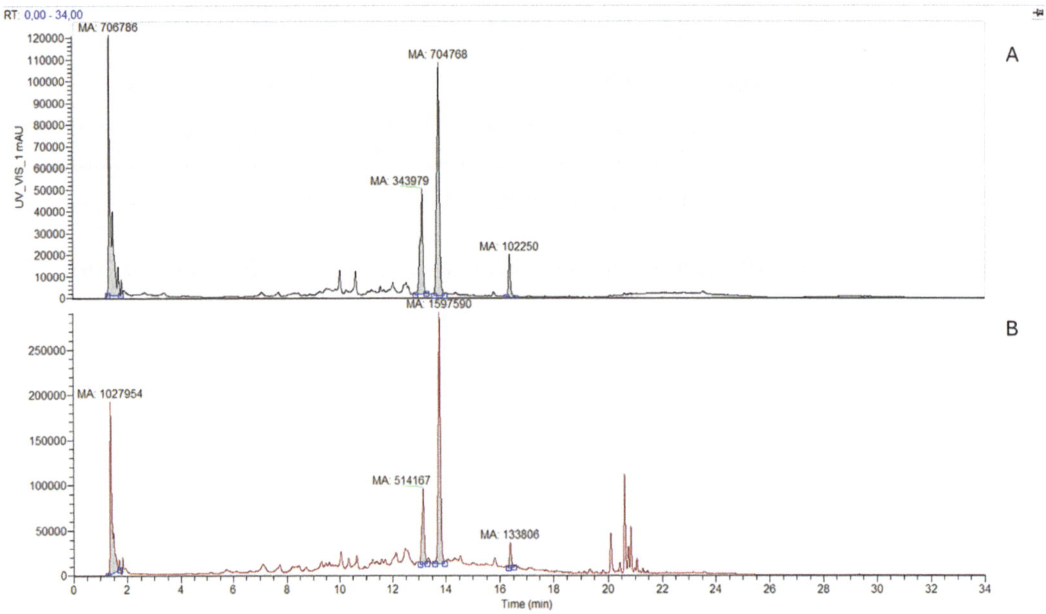

Fig. 2 Illustrative image showing possible variation that can be found in the polyphenol chromatographic profile between a control (**a**) sample and a sample with heavy metal-induced phytotoxicity (**b**), both recorded at 280 nm

Upon identification of the phenolic compounds present in the samples, chromatograms of control samples can be stacked with the chromatograms of heavy metal-induced phytotoxicity samples for comparison of the differences in terms of profiles and peak intensities as shown in Fig. 2. The presence of heavy metals will induce ROS and changes in the antioxidant systems of the plant, which will naturally translate into phenolic profile changes. These changes can be perceived as modifications in the peak areas, which indicate changes in the compounds' concentration, or appearance/disappearance of peaks, which indicate activation/deactivation of specific antioxidant defenses.

3.4 Quantification of the identified compounds

To gain deeper insight and enable more accurate information about the biochemical impact of the heavy metal-induced toxicity on plants' antioxidant defenses, estimation of the concentration of specific compounds can be carried out in control versus exposed samples. This can be achieved by quantification of the identified leaf compounds using calibration curves prepared with pure standard compounds. For that:

Prepare a range of concentrations of the pure compound (or representatives of each chemical family) at defined concentrations, usually ranging from 1 to 1000 μg mL^{-1} (*see* **Note 8**).

Inject each pure standard concentration following the same analytical protocol used for leaf sample analysis (chromatographic conditions described in Subheading 3.3).

Integrate the peaks of chromatograms obtained for each concentration of the standard compounds, and the corresponding compound in the samples/control to obtain the peak areas (*see* **Note 9**).

Plot the different concentrations of the standard compounds against the respective peak areas to obtain the equation of the curve ($y = mx + b$).

Substitute the y in the equation by the value of the area of the corresponding peak in the sample/control and calculate the value of the x, which will correspond to the concentration of the compound in the sample/control. The results can be expressed as the mass of a compound per mass of dry leaf tissue (i.e., the leaf tissue mass used in the extraction process; see Subheading 3.1) [e.g., 15–17].

Compare the concentrations of identified compounds between control and heavy metal-treated samples to obtain information to the extent to which heavy metal-induced toxicity affects their abundance.

Alternative method when standard compounds are unavailable—relative abundance approach:

Integrate the peak areas of a given compound eluting the control and samples.

Calculate the relative abundance of that compound using the formula [e.g., 12]:

$$\text{Relative peak area average } (\%) \\ = (\text{Compound peak area})/(\text{Total peak area}) \times 100$$

Compare the relative peak areas of identified compounds between control and heavy metal-treated samples to obtain information to the extent to which heavy metal-induced toxicity affects their abundance.

The final results can be expressed as mean (± standard deviation or error) of at least three independent analyses per treatment.

4 Notes

1. This protocol refers to the extraction of phenolic compounds from leaves but can also be used for other plant tissues. In leaves, the proportion of 1 g of leaf dry weight to 10 mL of solvent is used, for example, for stem the same proportion can be also used [18].
2. This material is only necessary if the semi-quantification of the identified compounds is carried out. Please see Subheading 3.4.
3. Due to the toxicity of methanol, always use nitrile gloves, as well as a lab coat and safety goggles with side protection. Also, all the procedures of metabolite extraction should be performed in a laboratory fume cupboard.
4. The choice of extraction solvent plays a crucial role in determining both the qualitative and quantitative composition of the extracted phenolic compounds. Commonly used solvents for extracting phenolics from leaves include methanol, ethanol, and acetone, among others. The optimal solvent depends on the specific phenolic compounds targeted and the characteristics of the plant material. Methanol is particularly effective for extracting low molecular weight polyphenols, such as flavonoids. Based on previous works [e.g., 11, 14–16], methanol was selected as the solvent for use in this protocol.
5. An alternative procedure for phenolic compounds extraction method is provided in case you have limited plant material availability. Also, this protocol can be performed with leaf fresh material [11].

6. The vial can be kept in the freezer for 2 weeks and then be used to inject in the chromatography equipment.
7. The samples can be dried to concentrate the compounds under a N2 flux or using SpeedVac equipment without heating.
8. The concentrations of the pure standard curves might need to be adjusted depending on the area of the peaks observable. Ideally, the area of the peak appearing in the sample should be in between the areas of the highest and lowest concentration of the pure standard compound.
9. Peak integration must be done to the maximum UV absorption of the identified compound. For example, flavonoids normally should be integrated on chromatograms recorded at 280 nm and for phenolic acids normally should be integrated on chromatograms recorded at 340 nm.

Acknowledgments

This work was supported by Fundação para a Ciência e Tecnologia (FCT), I.P., in the framework of the Projects UIDP/04004/2025 and UIDB/04004/2025—Centre for Functional Ecology—Science for the People & the Planet (DOI identifier 10.54499/UIDP/04004/2020). Thanks are due to the University of Aveiro, Fundação para a Ciência e Tecnologia, and Ministério da Ciência, Tecnologia e Ensino Superior (FCT/MCTES) for the financial support to LAQV-REQUIMTE research unit (project UIDP/50006). Susana M. Cardoso thanks FCT/MCTES for the funding provided through the Institutional Call to Scientific Employment Stimulus.

References

1. Lopez-Orenes A, Dias MC, Ferrer MA, Calderon A, Moutinho-Pereira J, Correia C, Santos C (2018) Different mechanisms of the metalliferous *Zygophyllum fabago* shoots and roots to cope with Pb toxicity. Environ Sci Pollut Res Int 25:1319–1330. https://doi.org/10.1007/s11356-017-0505-1
2. Oladoye PO, Olowe OM, Asemoloye MD (2022) Phytoremediation technology and food security impacts of heavy metal contaminated soils: a review of literature. Chemosphere 288. https://doi.org/10.1016/j.chemosphere.2021.132555
3. López-Orenes A, Santos C, Dias MC, Oliveira H, Ferrer MÁ, Calderón AA, Silva S (2021) Genotoxicity and cytotoxicity induced in *Zygophyllum fab*ago by low Pb doses depends on the population's redox plasticity. Horticulturae 7:455. https://doi.org/10.3390/horticulturae7110455
4. Dias MC, Mariz-Ponte N, Santos C (2019) Lead induces oxidative stress in *Pisum sativum* plants and changes the levels of phytohormones with antioxidant role. Plant Physiol Biochem 137:121–129. https://doi.org/10.1016/j.plaphy.2019.02.005
5. Silva S, Silva P, Oliveira H, Gaivão I, Matos M, Pinto-Carnide O, Santos C (2017) Pb low doses induced genotoxicity in *Lactuca sativa* plants. Plant Physiol Biochem 112:109–116. https://doi.org/10.1016/j.plaphy.2016.12.026
6. Mariz-Pontes N, Dias MC, Silva AMS, Santos C, Silva S (2021) Sub toxic levels of TiO_2-nanoparticles interact antagonistically with Al and Pb alleviating their toxicity. Plant

Physiol Biochem 167:1–10. https://doi.org/10.1016/j.plaphy.2021.07.021

7. Martin RE, Postiglione AE, Muday GK (2022) Reactive oxygen species function as signaling molecules in controlling plant development and hormonal responses. Curr Opin Plant Biol 69:102293. https://doi.org/10.1016/j.pbi.2022.102293
8. Anjitha KS, Sameena PP, Puthur JT (2021) Functional aspects of plant secondary metabolites in metal stress tolerance and their importance in pharmacology. Plant Stress 2:100038. https://doi.org/10.1016/j.stress.2021.100038
9. Ali B, Gill RA (2022) Editorial: heavy metal toxicity in plants: recent insights on physiological and molecular aspects. Front Plant Sci 13:1016257. https://doi.org/10.3389/fpls.2022.1016257
10. Dias MC, Pinto DCGA, Silva AMS (2021) Plant flavonoids: chemical characteristics and biological activity. Molecules 26:5377. https://doi.org/10.3390/molecules26175377
11. Khoddami A, Wilkes MA, Roberts TH (2013) Techniques for analysis of plant phenolic compounds. Molecules 18:2328–2375. https://doi.org/10.3390/molecules18022328
12. Gamito G, Monteiro CJP, Dias MC, Oliveira H, Silva AMS, Faustino MAF, Silva S (2024) Impact of Fe_3O_4-porphyrin hybrid nanoparticles on wheat: physiological and metabolic advance. J Hazard Mater 471:134243. https://doi.org/10.1016/j.jhazmat.2024.134243
13. Catarino MD, Alves-Silva JM, Falcão SI, Vilas-Boas M, Jordão M, Cardoso SM (2016) Chromatography as a tool for identification of bioactive compounds in honeybee products of botanical origin. In: Chemistry, biology and potential applications of honeybee plant-derived products. Bentham e-books, pp 89–149
14. Kalili KM, de Villiers A (2011) Recent developments in the HPLC separation of phenolic compounds. J Sep Sci 34:854–876. https://doi.org/10.1002/jssc.201000811
15. Dias MC, Pinto DCGA, Freitas H, Santos C, Silva MAS (2029) The antioxidant system in *Olea europaea* to enhanced UV-B radiation also depends on flavonoids and secoiridoids. Phytochemistry 170:112199. https://doi.org/10.1016/j.phytochem.2019.112199
16. Dias MC, Figueiredo C, Pinto DCGA, Freitas H, Santos C, Silva AMS (2019) Heat shock and UV-B episodes modulate olive leaves lipophilic and phenolic metabolite profiles. Ind Crop Prod 133:269–275. https://doi.org/10.1016/j.indcrop.2019.03.036
17. Piccini C, Cantini C, Cai G, Pinto DCGA, Silva AMS, Romi M, Dias MC (2022) Chemical profiling of two Italian *Olea europaea* (L.) varieties subjected to UV-B stress. Plants 11:680. https://doi.org/10.3390/plants11050680
18. Parri S, Cai G, Romi M, Cantini C, Pinto DCGA, Silva AMS, Dias MCP (2024) Comparative metabolomics of leaves and stems of three Italian olive cultivars under drought stress. Front Plant Sci 15:1408731. https://doi.org/10.3389/fpls.2024.1408731

Chapter 8

ELISA-Based Protocols for Mycotoxin Detection Across Different Matrices in Africa

Wilfred Angie Abia and Herman Philipe Nfombouot Njitoyap

Abstract

Accurate data on mycotoxin contamination and exposure patterns is crucial in contributing to food safety policies and public health interventions in low-resource nations in the developing world. In such settings, with inadequate research infrastructures and high-performance analytical tools, the enzyme-linked immunosorbent assay (ELISA) technique remains the most widely used technique to measure regulated mycotoxins (e.g., aflatoxins and fumonisins) in food, feed, blood, and urine samples. The results of ELISA technique alone or/in combination with food frequency intake survey data may generally be employed to estimate dietary exposure levels of populations and individuals. This report provides collection of ELISA-based mycotoxin detection protocols (including strengths and weaknesses) with guides for reasonable results.

Key words Africa, Food safety, ELISA, Sample analysis, Mycotoxins

1 Introduction

Mycotoxins are a structurally diverse group of secondary metabolites produced by toxigenic fungal species. To date, over 500 mycotoxins have been identified and characterized. Although mycotoxins generally have significant socioeconomic impacts and pose considerable health risks to humans and animals, only seven are currently considered major public health concerns. These "Big Seven" mycotoxins—aflatoxins, fumonisins, ochratoxins, deoxynivalenol, zearalenone, patulin, and citrinin—along with their metabolites (e.g., aflatoxin M1, a hydroxylated metabolite of aflatoxin B1), have been regulated to protect human and animal health.

For example, aflatoxins—produced by Aspergillus flavus and A. parasiticus, commonly found in feeds and foods such as corn and groundnuts—are known to be carcinogenic, genotoxic, mutagenic, and immunosuppressive.

Paula Alvito and Helena Oliveira (eds.), *Food Contaminants Toxicity*, Methods and Protocols in Food Science,
https://doi.org/10.1007/978-1-0716-5096-7_8,

Despite the socioeconomic and health risks posed by mycotoxins, measuring their levels in feed, food, urine, and blood remains a challenge, particularly in developing countries with limited research infrastructure. This has effectively limited the use of state-of-the-art, highly sensitive analytical systems such as liquid chromatography–tandem mass spectrometry (LC-MS/MS) and instead favored the use of relatively lower-sensitivity immunoassays, such as enzyme-linked immunosorbent assay (ELISA) techniques. ELISA uses antibodies to detect and quantify the presence of specific analytes, such as aflatoxins, in solid or liquid samples. It relies on the specific binding between an antibody and its corresponding antigen and utilizes enzymes to produce a measurable signal.

In resource-limited developing countries, where LC-MS/MS remains largely inaccessible, its use is further constrained by lengthy experimental procedures, the need for advanced analytical equipment, and the requirement for highly specialized technical expertise [1–3]. As a result, ELISA continues to gain ground, benefiting from the lack of access to LC-MS/MS, especially when simple, efficient, and user-friendly detection methods are needed to prevent mycotoxin contamination in products and to enable on-site evaluation. Broadly speaking, the development, optimization, and validation of immunoassay methods have been described in detail by Cox et al. [4].

Generally, mycotoxins are highly soluble in organic solvents such as methanol, acetonitrile, acetone, dichloromethane, and ethyl acetate but only sparingly soluble in water—except for fumonisins and patulin. Fumonisins are soluble in water and polar organic solvents such as methanol and acetonitrile but are insoluble in nonpolar solvents. Patulin is soluble in both water and a variety of organic solvents, including ethanol, methanol, and ethyl acetate. To enhance extraction, acidic buffers or water are frequently added to these organic solvents [5]. To green this approach, methanol can be substituted with ethanol or propan-2-ol, which are less toxic and biodegradable. For acetonitrile, ethyl acetate or a mixture of ethanol and water can be more effective and environmentally friendly options. Acetone can be replaced with ethanol or butanol, which is less harmful while still providing good extraction efficiencies. Instead of dichloromethane, consider using natural deep eutectic solvents (NADES) or bio-based solvents such as limonene, which have lower toxicity profiles. Lastly, ethyl acetate can be substituted with NADES or bio-based solvents such as 2-methyl tetrahydrofuran (2-MeTHF), offering effective extraction with reduced environmental impact [6, 7].

Since mycotoxins are nonantigenic, they are conjugated to carrier proteins (e.g., BSA) to generate antibodies. These techniques enable rapid, on-site testing with commercial kits widely used for aflatoxins, deoxynivalenol, and others. While less sensitive than chromatography, they balance practicality and cost, addressing acute and chronic toxicity risks in food and humans [8–12]. Ongoing developments aim to enhance cross-reactivity and sensitivity using nanomaterials and label-free approaches.

The classic competitive enzyme-linked immunosorbent assay (cELISA) and indirect enzyme-linked immunosorbent assay (indirect ELISA) are the most widely used immunochemical methods for mycotoxin analysis. In cELISA, the sample antigen is allowed to compete with the optimal concentration of the labeled antigen for a fixed amount of available binding sites in a microtiter plate pre-coated with mycotoxin-specific antibody [13, 14]. In contrast, in indirect ELISA, mycotoxin-conjugated proteins are reacted with the immobilized specific mycotoxin antibody and then with the developed labeled secondary antibody [15, 16]. The coloring reaction is started by adding the chromogenic solution-containing chromogen (such as TMB: 3,3′,5,5′-tetramethylbenzidine, and OPD: o-Phenylenediamine) to allow the color change. The reaction is stopped by the addition of a stop solution (commonly sulfuric acid at concentration from 0.5 to 2 M), and then the absorbance is measured by a microtiter plate reader [17–19]. If mycotoxin is present at a concentration in the medium or higher than its threshold value in a sample, the binding of enzyme-labeled antigen of mycotoxin and partner antibody to mycotoxin is released to react, and the resulting color formation is proportional to the mycotoxin concentration in the sample.

The methodology for detecting and semi-quantifying mycotoxins in biological specimens (blood, serum, and urine) and plant-based materials (feed and food) has been highlighted.

2 Materials

1. Follow the exact protocol as provided by the manufacturers. Prepare all solutions using distilled water and analytical grade reagents. Prepare and store all reagents at room temperature (unless indicated otherwise, e.g., prepare reagents just before use). Diligently follow all waste disposal regulations when disposing waste materials. Procedures for different matrices and reagents preparation will be described.

2.1 Sample Preparation

2.1.1 Blood/Serum Specimens

1. Serum separator tubes (SSTs) with clot activator.
2. Centrifuge compatible with blood collection tubes (capable of 3000× *g*).
3. Pipettes (adjustable, 100 μL capacity) and disposable tips.
4. Transfer tubes/pipettes for serum separation.
5. Protein saver cards (e.g., Whatman 903 filter paper).

6. Drying rack or clean surface for air-drying DBS/DSS.
7. Disposable gloves and lab coats.
8. Scissors or whole punch (for spot excision with ~1 mm margins).
9. Ruler/template (optional for consistent spotting).
10. 2 mL safe-lock microcentrifuge tubes.
11. Ultrasonic bath or probe sonicator.
12. Vacuum concentrator/centrifugal evaporator (e.g., SpeedVac) for drying.
13. Glassware/pipettes for solvent preparation (accurate measurement).
14. High-speed microcentrifuge (22,000× *g* capability).
15. Low-binding microcentrifuge tubes (1.5–2 mL).
16. Autosampler vials or clean microtubes for final supernatant.
17. Marker pens for labeling.
18. Tube racks and waste containers (for organic solvents/sharp objects).
19. Safety goggles and chemical fume hood (for handling volatile solvents).
20. Desiccant (optional for drying environment).
21. Reagents: HPLC-grade water, acetonitrile, and glacial acetic acid (95:5:0.1 v/v/v).

2.1.2 Urine Samples

1. Sterile, clean container (e.g., polypropylene cup) for collecting 10–50 mL urine.
2. Filter paper (optional, for dried urine spots as an alternative in field or home collection).
3. Refrigerator (for immediate storage if analysis is delayed).
4. Freezer (capable of −20 °C or lower for longer storage).
5. Dry ice (optional, for shipping frozen samples to labs).
6. Centrifuge (temperature-controlled, capable of 12,000× *g* at 4 °C).
7. 15 mL centrifuge tubes.
8. Vortex mixer.
9. Filter syringe.
10. Glass tubes (for storing filtrate at 4 °C).
11. Sodium chloride (NaCl).
12. Acetonitrile (ACN).
13. C18 sorbent.

2.1.3 Feed Samples

1. Ground feed (particle size <0.85 mm; stored at −20 °C if not analyzed immediately).
2. Analytical balance (for precise weighing).
3. 250 mL flask (for mixing).
4. Mechanical shaker (for agitation).
5. Funnel setup (for filtration).
6. Whatman No. 1 filter paper (for filtration).
7. Pipettes or graduated cylinder (for measuring liquids).
8. Labeling tape/markers (for sample identification).
9. Volumetric flasks or measuring cylinders (for precise solution preparation).
10. Gloves, lab coat, and safety goggles (for personal protection).
11. Waste containers (for disposal of solvents and used filter paper).
12. Vacuum pump (optional, to speed up filtration).
13. Reagents:
    - Methanol (HPLC-grade or equivalent purity).
    - Deionized or distilled water.
    - Methanol-water solution (70:30 v/v): Prepare by mixing 70 mL methanol with 30 mL water.

2.1.4 Tissue Samples

Solvents and Chemicals

1. Acetonitrile (ACN).
2. Methanol (MeOH).
3. Chloroform ($CHCl_3$).
4. Ultrapure water (from Milli-Q system).
5. Sodium chloride (NaCl).
6. Formic acid (HCOOH).
7. Ammonium formate ($HCOONH_4$).

Consumables

1. Tape/markers (for sample identification).
2. Syringe filters (0.2 μm × 13 mm, nylon).
3. Volumetric flasks or measuring cylinders (for precise solution preparation).
4. Gloves, laboratory coat, and safety goggles (for personal protection).
5. Waste containers (for disposal of solvents and used filter paper).

Samples (as materials for analysis)

1. Animal liver samples (pig, lamb, calf, chicken).

2.2 Reagents (Buffers) Preparation

2.2.1 Coating Buffer: 0.05 M Carbonate-Bicarbonate Buffer (pH 9.6)

Purpose:

1. Coating the microplate with AFM1-BSA.

Preparation:

1. Dissolve 1.59 g sodium carbonate (Na_2CO_3) and 2.93 g sodium bicarbonate ($NaHCO_3$) in 900 mL distilled water.
2. Adjust the pH to 9.6 using HCl or NaOH if necessary.
3. Bring the final volume to 1 L with distilled water.
4. Filter or sterilize if required. Store at 4 °C for short-term use.

2.2.2 Washing Buffer 1× PBS with 0.1% Tween 20 (PBS-T)

Purpose:

1. Washing steps to remove unbound reagents.

Preparation:

1. First, prepare 1× PBS (Phosphate-Buffered Saline):
   Dissolve in 1 L distilled water:
   - 8 g NaCl.
   - 0.2 g KCl.
   - 1.44 g Na_2HPO_4 (disodium hydrogen phosphate).
   - 0.24 g KH_2PO_4 (potassium dihydrogen phosphate).
   - Adjust pH to 7.4 using HCl or NaOH.
   - Autoclave or filter-sterilize for long-term storage.
2. To make PBS-T (washing buffer):
   - Add 1 mL Tween 20 (0.1% final concentration) to 1 L of 1× PBS,
   - Mix thoroughly and store at room temperature.

2.2.3 Blocking Buffer: 5% Skim Milk in PBS-T

Purpose:

1. Blocking nonspecific binding sites on the microplate.

Preparation:

1. Add 5 g skim milk powder to 100 mL PBS-T (prepared as above).
2. Stir or vortex until fully dissolved.
3. Centrifuge briefly to remove undissolved particles (optional).

   Note: Prepare fresh before use to avoid contamination.

2.2.4 Dilution Buffer for HRP Conjugate: 5% Skim Milk in PBS-T

Purpose:

1. Diluting the anti-His HRP antibody (same as blocking buffer).

   Preparation:
   Follow the same steps as the blocking buffer (5% skim milk in PBS-T).

2.2.5 Stopping Solution 1 M H_2SO_4

Purpose:

1. Stopping the TMB enzymatic reaction (not a buffer, but included for completeness).

Preparation:

1. Handle concentrated sulfuric acid (H_2SO_4) with gloves and eye protection.
2. Slowly add 55.5 mL of concentrated H_2SO_4 (18 M) to 944.5 mL distilled water (always add acid to water, not vice versa).
3. Mix carefully and store in a labeled, acid-resistant container.

3 Methods

3.1 Collection and Mycotoxins Extraction Protocols

3.1.1 Serum from Blood Samples (Adapted from [20]) Collection

1. Collect whole blood in EDTA (Ethylene Diamine Tetraacetic Acid) blood collection tubes.

Extraction:

1. Centrifuge the blood specimen at 3000× *g* for 10 min (to separate the plasma, which will be the supernatant, from the buffy coat for further processing).
2. Mix in glass test tubes 100 μL of plasma and 150 μL ethyl acetate, then vortex for 20 min.
3. Centrifuge at 25,830× *g* (4 °C, 10 min).
4. Collect organic layer (70 μL after first step, 100 μL after second, 200 μL after 3rd).
5. Evaporate to dryness, reconstitute in 400 μL 20% methanol.

3.1.2 Dried Blood Spots (DBS) [21]

1. Dried blood spots (DBS) are a common microsampling technique used for collecting, storing, and analyzing small volumes of blood, often on filter paper cards. They are widely used in newborn screening, diagnostic testing, and research due to their stability and ease of transport. The process typically involves the following steps:

Collection:

1. Blood Collection: Obtain a small amount of capillary blood via a finger prick (for adults or older children) or heel stick (for infants). Alternatively, venous blood can be collected through venipuncture and then spotted.
2. Spotting the Blood: Apply 1–2 drops (approximately 50–75 μL per spot) of blood to the center of pre-printed circles on absorbent filter paper (e.g., Whatman 903 cards). Ensure the blood fully saturates each circle without touching the paper directly to avoid contamination. Typically, 4–5 spots are collected per card.
3. Drying: Allow the spots to air-dry completely at room temperature for at least 3–4 h (or overnight) in a clean, well-ventilated area. Prop the card up to avoid contact with surfaces, and protect it from direct sunlight, humidity, heat, insects, or contaminants. Do not stack cards while drying.
4. Storage and Transportation: Once dry, place the card in a protective envelope or bag with a desiccant packet to maintain low humidity. Store at room temperature for short-term use or refrigerate/freeze for longer periods. Cards can be shipped via regular mail without biohazard concerns if properly packaged.

Extraction:

1. Spot Excision: DBS/DSS spots were cut out, leaving ~1 mm margins around the visible spot edge.
2. Solvent Extraction: Spots were placed in 2 mL safe-lock tubes and extracted with 1 mL of a water/acetone/acetonitrile mixture (30:35:35, v/v/v) under 30 min of sonication.
3. Evaporation: 800 μL of the extract was transferred to a new tube and evaporated to dryness at 50 °C under reduced pressure.
4. Residues were reconstituted in 95:5:0.1 (v/v/v) water/acetonitrile/acetic acid.
5. The mixture was centrifuged at 22,000× *g* for 10 min, and supernatant will transfer in new vial tubes.

3.1.3 Dried Serum Spots (DSS) [21]

1. Dried Serum Spots (DSS), a similar technique as DBS but using serum (the liquid component of blood after clotting and centrifugation) instead of whole blood. The process involves an additional step to isolate serum and is as follows:

Collection:

1. Blood Collection and Serum Separation: Collect venous blood (typically 5–10 mL) in a tube without anticoagulants to allow clotting. Centrifuge the sample (e.g., at 3000–5000 g for

10–15 min) to separate the serum from the cellular components.

2. Spotting the Serum: Pipette or drop 50–150 μL of the separated serum onto absorbent filter paper (similar to DBS cards, often prepunched discs). Blot a large drop to form a spot approximately 30–35 mm in diameter, ensuring even absorption without oversaturation.
3. Drying: Air-dry the spots at room temperature for at least 5 h (or overnight) in a clean, dry environment. Similar to DBS, avoid direct contact with surfaces and protect from humidity or contaminants.
4. Storage and Transportation: Store the dried spots in sealed bags with desiccants at −20 °C or − 80 °C for long-term stability.

Extraction:

1. Extraction is done as for DBS.

3.1.4 Urine Collection [22, 23]

1. Sample Type and Timing: Collect a random spot urine sample, preferably the first morning void for higher concentration of metabolites. No fasting is strictly required, but fasting for 12 h prior may enhance mycotoxin excretion in some cases. Avoid excessive hydration (e.g., no more than 8 oz of water in the hour before collection) to prevent dilution.
2. Volume and Container: A volume of 10–50 mL is usually sufficient. Use a sterile, clean container (e.g., polypropylene cup) to avoid contamination. For field or home collection, dried urine spots on filter paper can be an alternative, requiring minimal equipment and suitable for remote areas.
3. Storage and Transport: Refrigerate the sample immediately after collection if analysis is delayed. For longer storage, freeze at −20 °C or lower to preserve stability. Ship frozen samples to labs using dry ice if needed. Commercial kits often include detailed instructions for handling. Creatinine levels are often measured post-collection to normalize mycotoxin concentrations for urine dilution variability.

Extraction [24]:

Mycotoxin Extraction Protocols Using Salting-Out Liquid-Liquid Extraction (SALLE)

1. Centrifuge 1 mL urine at 10,621× *g* for 3 min at 4 °C; transfer supernatant to a 15 mL tube.
2. Add 0.3 g NaCl, vortex to mix.
3. Add 1 mL ACN and 30 mg C18 sorbent; vortex for 1 min.

4. Centrifuge at 4500× *g* for 3 min at 4 °C.
5. Filter the upper layer using filter syringe.

   Store filtrate on glass tubes at 4c until analysis.

3.1.5 Feed (Adapted from [19]) Collection

1. Mill or ground feed using crusher to size <0.85 mm and store at -20C if analysis will not run immediately.

Extraction:

1. Accurately measure 20 g of ground feed using an analytical balance and transfer it to a 250 mL flask.
2. Add 100 mL of a methanol:water solution (70:30 ratio, v/v) into the flask.
3. To prepare 100 mL of solution, mix 70 mL of methanol with 30 mL of water, shake gently.
4. Secure the flask on a mechanical shaker and agitate vigorously for 3 min.
5. Allow the mixture to stand undisturbed at room temperature for 5 min.
6. Filter the supernatant through Whatman No. 1 filter paper to separate particulates.
7. Collect 10 mL of the filtrate and mix it with 50 mL of fresh methanol:water solution (70:30, v/v) for further analysis.

3.2 Green Protocols for Mycotoxins Extraction

3.2.1 Food (Adapted from [25]) Collection

1. Collect and accurately weigh 1.0 g of contaminated food, such as maize, groundnuts, soyabeans, rice or other cereals and nuts, sample into a 15 mL polypropylene tube.

Green Extraction:

1. Extraction: Add 5 mL of extraction solution (0.1% PEG 20 K, 20% ethanol in 10 mM acetate buffer, pH 5.0) to the tube.
2. Mixing: Manually vortex the mixture for 1 min at medium intensity to ensure thorough homogenization.
3. Filtration: Filter the mixture through a 0.45 μm membrane filter into a fresh Eppendorf tube to remove particulates.
4. Dilution: Combine the filtrate with an equal volume of running buffer (3.9% PEG 20 K, 2% Tween 20, 1% BSA in 10 mM acetate buffer, pH 5.0).
5. Final Mixing: Briefly vortex the diluted solution to ensure complete mixing before analysis.

3.2.2 Tissue [26] Collection

1. Collect an appropriate tissue, such as liver, without any additives in an ice cooler container and transport to the laboratory for analysis.

Green Extraction:

1. SALLE (Salting-Out Assisted Liquid-Liquid Extraction):
   - Sample Preparation: Add 3 g of tissue (e.g., liver) sample to a 15 mL tube.
   - Hydration: Mix with 3 mL of ultrapure water and homogenize by vortexing for 10 seconds.
   - Solvent Addition: Add 3 mL of acetonitrile (ACN) to the tube.
   - Extraction: Vortex the mixture vigorously for 2 min to ensure thorough mixing.
   - Salting Out: Add 1 g of sodium chloride (NaCl), shake manually for 1 min, and then centrifuge at 4500 rpm for 5 min to separate phases.
2. Supernatant Processing:
   - Filtration: Collect ~2 mL of the upper supernatant layer and filter through a 0.2 μm nylon membrane filter into a clean vial.
3. DLLME (Dispersive Liquid-Liquid Microextraction):
   - Extractant Addition: Mix the filtered supernatant with 600 μL of chloroform ($CHCl_3$).
   - Dispersion: Quickly transfer this mixture into a tube containing 6 mL of ultrapure water, forming a cloudy solution (emulsification).
   - Phase Separation: Centrifuge at 3500 rpm for 3 min to sediment the chloroform phase.
4. Post-extraction Handling:
   - Sediment Collection: Use a syringe to collect the lower chloroform phase (sedimented layer).
   - Drying: Evaporate the chloroform under a gentle airstream until dry.
5. Reconstitution:
   - Solubilization: Redissolve the dried residue in 500 μL of methanol-water (50:50, v/v).
   - Final Filtration: Filter the solution through a 0.2 μm membrane filter to ensure clarity.

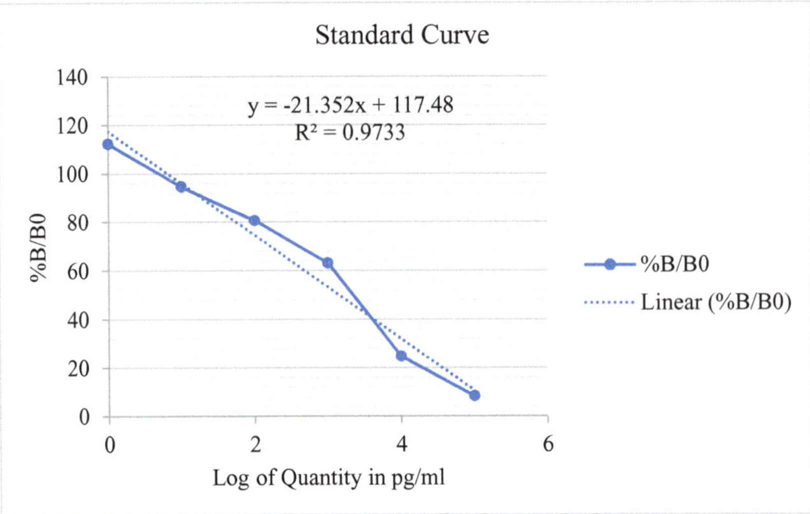

Fig. 1 An example of a standard curve for ELISA-based mycotoxin analysis

3.3 ELISA Protocols for the Detection of Mycotoxins

3.3.1 In House Competitive Indirect ELISA for Aflatoxin Determination (Adapted from [18]) Preparation of Plates

1. Coating
   - Coat a microplate with 100 μL of 2 μg/mL AFM1-BSA in 0.05 M carbonate buffer (pH 9.6).
   - Incubate the plate overnight at 4 °C.
2. Washing (Post-coating): Wash the plate three times with 300 μL of washing buffer per well.
3. Blocking:
   - Add 300 μL of blocking buffer (5% skim milk in 1× PBS with 0.1% Tween 20) to each well.
   - Incubate at 37 °C for 1 h.
4. Washing (Post-blocking): Repeat the washing step three times as described above.

Preparing Mycotoxin Standards and samples

1. Antibody, samples and AFM1 Standard Incubation:
   - Add 50 μL of samples, 50 μL of serial dilutions of AFM1 standard followed by 50 μL of antibodies anti-AFM1-BSA to each well.
   - Incubate the plate at 37 °C for 1 h.
2. Washing (Post-Antibody): Wash the plate four times with 300 μL of washing buffer per well.

3. HRP Conjugate Addition:
   - Add 100 μL of anti-His tag HRP monoclonal antibody (diluted 1:10,000 in 5% skim milk) to each well.
   - Incubate at 37 °C for 1 h.
4. Final Washing: Wash the plate five times with 300 μL of washing buffer per well.
5. TMB Substrate Reaction:
   - Add 100 μL of TMB solution to each well and incubate for 7 min at room temperature.
   - Stop the enzymatic reaction by adding 50 μL of 1 M H_2SO_4.
6. Measurement: Measure the absorbance at 450 nm using a microplate reader.

Data Analysis

1. Fit the standard curve using the absorbance data as shown in Fig. 1.
2. Calculate B/B_o, where B represents the absorbance in the presence of AFM1 standard, and B_o represents the absorbance in its absence.
3. Determine the IC_{50}, defined as the AFM1 concentration that inhibits 50% of antibody binding to the coated antigen. This can be done by plotting the percentage of inhibition versus the log concentration of the inhibitor and then using a nonlinear regression curve-fitting program to determine the IC50 value. Use a nonlinear regression curve-fitting software (like GraphPad Prism, R, or similar) to fit the data. A common model used is the four-parameter logistic (4PL) model.

3.3.2 In House Competitive Indirect ELISA for AFB1, DON, FUM, OTA, and ZEA, Determination (Adapted from [14])

Preparation of plates

1. Coating the Plates: Coat ELISA microplates with 100 μL of protein G at a concentration of 2500 ng/mL (ppb) per well in 100 mM carbonate buffer and incubate the plates at room temperature (RT) for 1 h.
2. Washing the Plates: Wash the plates five times with 200 μL of 1 × PBST.
3. Coating with Monoclonal Antibodies: Immediately coat the plates with 100 μL per well of monoclonal antibodies against AFB1 (625 ppb), DON (2500 ppb), FUM (2500 ppb), OTA (625 ppb), and ZEA (1250 ppb), diluted in 1 × PBS, and incubate at RT for 1 h.
4. Washing Again: Wash the plates five times with 200 μL of 1 × PBS.

5. Blocking/Stabilizing: Block/stabilize the plates with 300 μL of blocking buffer and incubate for 1 h at RT.
6. Final Wash and Drying: Wash the plates once with 300 μL of double-distilled water and air-dry the plates for 1 h at RT. Store the plates at 4 °C in a sealed foil wrap with desiccant.

Preparing Mycotoxin Standards and Samples

Use a 2% BSA in 1× PBS as the diluent for all mycotoxin standards and as a surrogate for the human matrix.

1. Dissolve mycotoxin standards in 2% BSA/PBS using a four-fold serial dilution:
2. DON: 1250 to 0.076 ppb,
3. AFB1 and ZEA: 625 to 0.038 ppb,
4. FUM B1/B2: 312.5 to 0.019 ppb.
5. Preparing HRP Conjugates: Prepare HRP-conjugated AFB1 (312.5 ppb), DON (625 ppb), FUM (156.25 ppb), OTA (156.25 ppb), and ZEA (312.5 ppb) using the Guardian™ reagent.
6. Adding Standards and HRP Conjugates: Introduce 50 μL of AFB1 standard in 2% BSA/PBS to microwells coated with a monoclonal antibody against AFB1.
7. Add 50 μL of AFB1-HRP in Guardian™ reagent to the same wells.
8. Repeat this procedure for DON, FUM, OTA, and ZEA.
9. Incubation: Incubate the plates for 1 h at RT on a plate shaker at 500 rpm.
10. Washing: Wash the plates five times with 200 μL of 1 × PBS.
11. Adding TMB: Supplement the plates with 100 μL of TMB and incubate for 30 min in the dark at RT.
12. Stopping the Reaction: Stop the catalytic reaction by adding 100 μL of 2 M sulfuric acid.
13. Measuring Optical Density: Measure the optical density (OD) at 450 nm.

Data Analysis

1. Employ the four-parameter logistic (4PL) and Gaussian fits to establish a nonlinear relationship between mycotoxin concentrations in 2% BSA/PBS (x-axis) and OD at 450 nm (y-axis).
2. Include background correction, log10 transformation of mycotoxin concentrations, and signal-to-noise (%B/B0) assessment as shown in Fig. 1.

3. Fit replicated data points with 4PL so that the curve's upper and lower asymptotes equal B and B0 (0 and 100%).
4. Use Gaussian fitting to include the average %B/B0 from the replicates on the ordinate scale.

4 Notes

1. Mycotoxin extraction from blood: Dried blood spots (DBS) and dried serum spots (DSS) offer distinct advantages and limitations. DBS enable minimally invasive sampling, requiring only small blood volumes (ideal for infants, animals, or remote settings), simplified storage/shipping, and cost efficiency, though they face challenges like hematocrit variability, matrix interference, and analyte stability [27]. DSS, while advantageous for serum-specific assays (e.g., immunological tests) and reduced matrix complexity, require prior serum separation via centrifugation and contend with larger spot sizes due to lower viscosity, complicating volume quantification. Both methods share logistical benefits (room-temperature stability, low volume needs) but suffer from standardization issues and potential analyte degradation. While DBS are widely adopted for biomonitoring and clinical applications, DSS remain less standardized, emphasizing trade-offs between practicality (ease of use, cost) and analytical precision, necessitating tailored validation for specific contexts [21, 27].

   The strengths of the method include its minimally invasive nature, making it suitable for vulnerable populations such as infants and animals, as well as for use in resource-limited settings; the stability of dried blood spots (DBS) and dried serum spots (DSS), which allow for room-temperature storage and reduce logistical burdens; and its cost-effectiveness, as it relies on simple materials like filter paper and basic solvents. However, there are limitations such as hematocrit variability in DBS, which can affect spot size and volume consistency; matrix interference from hemoglobin in DBS, which may compromise analytical accuracy; and the risk of analyte degradation during drying and storage, particularly for light-sensitive mycotoxins. Key considerations include ensuring complete coagulation of serum to avoid fibrin interference, using a water/acetone/acetonitrile (30:35:35) extraction solvent to balance polarity for broad mycotoxin recovery, and employing a reconstitution buffer of 95:5:0.1 water/acetonitrile/acetic acid to stabilize acid-labile mycotoxins like DON and ZEN.
2. Mycotoxin extraction from feed: The method boasts several strengths, including its high-throughput capability, making it suitable for processing large feed samples of 20 g, and its

simplicity, as the methanol-water (70:30) solution is cost-effective and efficiently extracts both polar and non-polar mycotoxins. However, limitations arise from matrix complexity, as various feed components, such as fibers and fats, may co-extract alongside mycotoxins, necessitating additional cleanup steps. Key notes for implementation include the preparation of the methanol-water solution by mixing 70 mL of HPLC-grade methanol with 30 mL of ultrapure water, while emphasizing safety precautions due to the toxicity of methanol, which should be handled in a fume hood. Additionally, storing ground feed at -20 °C is crucial to prevent mycotoxin degradation and microbial growth.

3. Mycotoxin extraction from tissue: This method offers notable strengths, including a dual-phase extraction combining acetonitrile (ACN) for protein precipitation and dispersive liquid-liquid microextraction (DLLME) for mycotoxin enrichment, alongside high recovery rates due to chloroform's efficiency in extracting hydrophobic mycotoxins like aflatoxin B1 (AFB1) and zearalenone (ZEN). However, limitations include chloroform toxicity, necessitating strict safety protocols (gloves, fume hood), and a labor-intensive workflow involving multiple centrifugation and filtration steps. Key aspects include the ACN/-water/NaCl mixture to ensure protein precipitation and phase separation and a reconstitution solvent (MeOH:H_2O, 50:50) optimized for LC-MS compatibility. Critical operational notes emphasize solvent compatibility (use glass or chemical-resistant polypropylene tubes to avoid plastic degradation by chloroform), safety measures (handle volatile/toxic chloroform and ACN in a fume hood with PPE), HPLC-grade solvent purity to prevent analytical interference, and centrifuge tube specifications (rated for high-speed use and solvent resistance) to ensure procedural integrity [8, 24].
4. Mycotoxin extraction from tissue using a green sample approach: This method offers strengths such as reduced solvent use through the integration of salting-out assisted liquid-liquid extraction (SALLE) with acetonitrile (ACN) and sodium chloride (NaCl) and dispersive liquid-liquid micro extraction (DLLME) using chloroform ($CHCl_3$), alongside improved selectivity as SALLE minimizes matrix interference prior to DLLME enrichment. However, limitations include the risk of emulsion formation during DLLME, which can compromise phase separation, and manual syringe-based phase collection, introducing variability. Key operational notes emphasize the SALLE buffer (ACN/NaCl) requiring precise NaCl addition for phase separation, HPLC-grade chloroform for DLLME to avoid impurities, and glass tubes for chloroform steps to prevent plastic degradation. Strict safety protocols (PPE, fume

Table 1
Selected organic solvents compatible with mycotoxin extraction and ELISA detection

Mycotoxin	Extraction solvent	ELISA-compatible prep
Aflatoxins	MeOH:H_2O (70:30)	Dilute 1:10 in PBS + 1% BSA
Ochratoxin A	ACN:H_2O (60:40)	Evaporate, reconstitute in PBS/Tween-20
Fumonisins	ACN:MeOH:H_2O (25:25:50)	Dilute 1:5 + SPE cleanup (use *C18, immunoaffinity, or MycoSep® columns* to transfer mycotoxins into aqueous phase)
Zearalenone	EtOH:H_2O (80:20)	Dilute 1:8 in buffer (max 10% EtOH)

hood) are mandatory due to chloroform's toxicity/carcinogenicity and ACN's flammability. All solvents must be HPLC-grade to ensure LC-MS compatibility, while centrifuge tubes must tolerate specified speeds and solvent exposure. Gentle evaporation (air/nitrogen streams) is critical to avoid overheating and preserve analyte integrity during solvent removal [8, 26].

5. General extraction considerations: Selecting organic solvents compatible with both efficient mycotoxin extraction and downstream ELISA detection requires careful balancing, as most solvents that extract mycotoxins effectively can interfere with antibody binding. No organic solvent is directly compatible with ELISA antibodies at extraction concentrations. Use dilution, evaporation, or clean-up to bridge the gap. Methanol ≤10% and ethanol ≤20% are most manageable post-dilution. Solvents >10–20% can denature antibodies or block antigen-antibody binding [28, 29]. Sometimes quenching/neutralization are needed before analysis as shown in Table 1.

6. ELISA: In ELISA protocol, all incubation steps are conducted at 37 °C unless otherwise specified, with consistent washing between steps using a PBS (1×)-based washing buffer containing Tween 20 to minimize nonspecific binding and background noise. The competitive binding efficiency between free aflatoxin M1 (AFM1) and the immobilized AFM1-BSA antigen is quantified via the B/B_0 ratio, reflecting the displacement of antibodies by free toxin. PBS serves as the base for preparing all buffers, including washing, blocking (supplemented with skim milk to block nonspecific protein interactions), and antibody dilution buffers. Buffers should be prepared with high-quality reagents and distilled/deionized water, adjusted to optimal pH to ensure binding efficacy, and stored at 4 °C (except stopping solution) for short-term use, with regular

Table 2
Positions of blanks, mycotoxin standards, and samples in ELISA plate

	1	2	3	4	5	6	7	8	9	10	11	12
A	B	S 2	S 10	S 18	S 26	S 34	B	S 2	S 10	S 18	S 26	S 34
B	STD 1	S 3	S 11	S 19	S 27	S 35	STD 1	S 3	S 11	S 19	S 27	S 35
C	STD 2	S 4	S 12	S 20	S 28	S 36	STD 2	S 4	S 12	S 20	S 28	S 36
D	STD 3	S 5	S 13	S 21	S 29	S 37	STD 3	S 5	S 13	S 21	S 29	S 37
E	STD 4	S 6	S 14	S 22	S 30	S 38	STD 4	S 6	S 14	S 22	S 30	S 38
F	STD 5	S 7	S 15	S 23	S 31	S 39	STD 5	S 7	S 15	S 23	S 31	S 39
G	STD 6	S 8	S 16	S 24	S 32	S 40	STD 6	S 8	S 16	S 24	S 32	S 40
H	S 1	S 9	S 17	S 25	S 33	S 41	S 1	S 9	S 17	S 25	S 33	S 41

contamination checks. Proper storage and reagent integrity are critical to maintaining assay reliability, ensuring accurate results while minimizing experimental variability.

ELISA tests for the detection of mycotoxins are mostly competitive ELISAs. In these tests, standards and samples are added followed by an enzyme-linked competitive mycotoxin. The interaction between the capture antibody or antigen and the mycotoxin in the sample generally begins before the addition of the competitive antigen or antibody. The amount of synthetic mycotoxin (antigen) or antibodies bound to the immobilized antibody is inversely proportional to the concentration of free mycotoxin in the sample. This generally leads to an overestimation of the mycotoxin concentration in the sample. In order to solve this problem, we use one of the following two methods:

Samples and standards are not added in duplicate at the same time in the plate, i.e., samples and standards are added first in the first half of the plate and then the operation is repeated in the other half, as shown in Table 2. Finally, the sample concentrations and the optical densities of standards and samples are obtained from the average of the optical densities of samples and standards obtained from the first part of the plate and the second. For example: mean of Blank OD = (blank A1 + blank A7)/2.

In a glass test tube, a mixture of the sample or the standard and the detection antigen or antibodies is made in equal volumes and then deposited in duplicate in the wells of the microtiter plate. In this method, the competition for the binding of the capture antibody or antigen begins at the same time for mycotoxin in sample and antigen or antibody respectively and does not allow an overestimation of the mycotoxin concentration in the sample. The average of the optical densities can easily be used for the calculation.

References

1. Bessaire T, Mujahid C, Mottier P, Desmarchelier A (2019) Multiple mycotoxins determination in food by LC-MS/MS: an international collaborative study. Toxins (Basel) 11(11):658
2. Maragos CM, Busman M (2010) Rapid and advanced tools for mycotoxin analysis: a review. Food Addit Contam Part A Chem Anal Control Expo Risk Assess 27(5):688–700
3. Wolf K, Schweigert FJ (2018) Mycotoxin analysis: a focus on rapid methods. Partnership for aflatoxin control in Africa. 1–46 pp. Available from: http://www.aflatoxinpartnership.org/
4. Cox KL, Devanarayan V, Kriauciunas A, Manetta J, Montrose C, Sittampalam S (2019) Immunoassay methods immunoassay development, optimization and validation flow chart 2 assay guidance manual; (Md):1–39. Available from: https://www.ncbi.nlm.nih.gov/books/
5. Magoke GZ, Alders RG, Krockenberger M, Bryden LW (2016) Aflatoxin and mycotoxin analysis: an overview including options for resource-limited settings. Intech;11(tourism):13. Available from: https://doi.org/10.1039/C7RA00172J, https://www.intechopen.com/books/advanced-biometric-technologies/liveness-detection-in-biometrics
6. Pradanas-gonz F, Angel M, Andrade-bartolom E, Navarro-villoslada F, Benito-pe E (2023) Extracting mycotoxins from edible vegetable oils by using green, ecofriendly deep eutectic solvents. Food Chem 429:136846
7. Bian Y, Zhang Y, Zhou Y, Wei B, Feng X (2023) Recent insights into sample pretreatment methods for mycotoxins in different food matrices: a critical review on novel materials. Toxins 15:215
8. Castell A, Arroyo N, Rosa M, Manrique P, Campillo N, Torres C (2024) Evaluation of distribution of emerging mycotoxins in human tissues : applications of dispersive liquid – liquid microextraction and liquid chromatography – mass spectrometry. Anal Bioanal Chem 416(2):449–459. https://doi.org/10.1007/s00216-023-05040-8
9. Oplatowska-Stachowiak M, Reiring C, Sajic N, Haasnoot W, Brabet C, Campbell K et al (2018) Development and in-house validation of a rapid and simple to use ELISA for the detection and measurement of the mycotoxin sterigmatocystin. Anal Bioanal Chem 410(12):3017–3023
10. Bowers E, Hellmich R, Munkvold G (2014) Comparison of fumonisin contamination using HPLC and ELISA methods in bt and near-isogenic maize hybrids infested with European corn borer or western bean cutworm. J Agric Food Chem 62(27):6463–6472
11. Peng H, Chang Y, Baker RC, Zhang G (2020) Interference of mycotoxin binders with ELISA, HPLC and LC-MS/MS analysis of aflatoxins in maize and maize gluten. Food Addit Contam Part A Chem Anal Control Expo Risk Assess 37(3):496–506
12. Kos J, Janić Hajnal E, Jajić I, Krstović S, Mastilović J, Šarić B et al (2016) Comparison of ELISA, HPLC-FLD and HPLC-MS/MS methods for determination of aflatoxin M1 in natural contaminated Milk samples. Acta Chim Slov 63(4):747–756
13. Alnaemi HS, Dawood TN, Algwari QT (2023) Plasma-activated water application for detoxification of aflatoxin B1, ochratoxin A, and fumonisin B1 in poultry feeds. Open Vet J 13(12):1654–1668
14. Garg K, Villavicencio-Aguilar F, Solano-Rivera F, Gilbert L (2022) Analytical validation of a direct competitive ELISA for multiple mycotoxin detection in human serum. Toxins (Basel) 14(11):727
15. Wang Y, Wang H, Li P, Zhang Q, Kim HJ, Gee SJ et al (2013) Phage-displayed peptide that mimics aflatoxins and its application in immunoassay. J Agric Food Chem 61(10):2426–2433
16. Tian J, Luo J, Qin J, Wang Y, Sun X, Zhang J et al (2023) Preparation of a broad-specificity antibody against zearalenone and its primary analogues and development of immunoassay of Coicis Semen and related products. J Food Sci 88(6):2723–2734
17. Ling S, Pang J, Yu J, Wang R, Liu L, Ma Y et al (2014) Preparation and identification of monoclonal antibody against fumonisin B (1) and development of detection by Ic-ELISA. Toxicon 80:64–72

18. Yi L, Liu H, Liu Y, Aiyisi, He J, Ming L et al (2025) Nanobody-based indirect competitive ELISA for the detection of aflatoxin M1 in dairy products. Sci Rep 15(1):785. https://doi.org/10.1038/s41598-024-83869-4
19. Merrill BE, Matson RS (2023) Mycotoxin quantification by competitive ELISA. Methods Mol Biol 2612:183–194
20. Slobodchikova I, Vuckovic D (2018) Liquid chromatography – high resolution mass spectrometry method for monitoring of 17 mycotoxins in human plasma for exposure studies. J Chromatogr A 1548:51–63. https://doi.org/10.1016/j.chroma.2018.03.030
21. Osteresch B, Viegas S, Cramer B, Humpf HU (2017) Multi-mycotoxin analysis using dried blood spots and dried serum spots. Anal Bioanal Chem 409(13):3369–3382
22. Aly AA, Górecki T (2020) Green approaches to sample preparation based on extraction techniques. Molecules 25(7):1–33
23. Schmidt J, Lindemann V, Olsen M, Cramer B, Humpf HU (2021) Dried urine spots as sampling technique for multi-mycotoxin analysis in human urine. Mycotoxin Res 37(2):129–140. https://doi.org/10.1007/s12550-021-00423-1
24. Escrivá L, Manyes L, Font G, Berrada H (2017) Mycotoxin analysis of human urine by LC-MS/MS: a comparative extraction study. Toxins (Basel) 9(10):330
25. Jintamethasawat R, Doljirapisit N, Somboonkaew A, Noppakuadrittidej P, Petchkongkaew A, Karoonuthaisiri N (2024) A multiplex microarray lateral fl ow immunoassay device for simultaneous determination of fi ve mycotoxins in rice. npj Sci Food. https://doi.org/10.1038/s41538-024-00342-2
26. Castell A, Arroyo-Manzanares N, Campillo N, Torres C, Fenoll J, Viñas P (2023) Bioaccumulation of mycotoxins in human forensic liver and animal liver samples using a green sample treatment. Microchem J 185:108192. Available from: https://www.sciencedirect.com/science/article/pii/S0026265X22010207
27. Lauwers M, Croubels S, De Baere S, Sevastiyanova M, Sierra EMR, Letor B et al (2019) Assessment of dried blood spots for multi-mycotoxin biomarker analysis in pigs and broiler chickens. Toxins (Basel) 11(9):541
28. Urusov AE, Zherdev AV, Petrakova AV, Sadykhov EG, Koroleva OV, Dzantiev BB (2015) Rapid multipleimmunoenzyme assay of mycotoxins. Toxins (Basel) 7(2):238–54
29. Lou X, Zhu A, Luo Q, Zhang Y, Long F (2017) Effects of organic solvents on immunosensing assays for small molecules based on an optofluidic immunosensing platform. Anal Methods [Internet] 9(39):5731–40. https://doi.org/10.1039/C6AY03373C

Part II

In Vitro & Ex Vivo Toxicological Approaches

Chapter 9

From Protocol to Analysis: Performing and Interpreting the MTT Assay

Andreia Rego, José Furtado, José A. Silva, Paula Alvito, and Carla Motta

Abstract

The 3-(4,5-dimethylthiazol-2-yl)-2,5-diphenyltetrazolium bromide (MTT) assay is a widely used method to assess cell metabolic activity as an indicator of cell viability, proliferation, and cytotoxicity. The assay is based on reducing the tetrazolium salt MTT to formazan by mitochondrial enzymes in living cells. This chapter describes the detailed protocol for performing the MTT assay, including preparation of reagents, cell plating, MTT incubation, and formazan solubilization. Key considerations and troubleshooting tips are also discussed. Additionally, available tools for calculating IC_{50} are also indicated.

Key words MTT assay, Cell viability, Cytotoxicity, Metabolic activity, IC_{50}

1 Introduction

First described by Mosmann in 1983 [1], the 3-(4,5-dimethylthiazol-2-yl)-2,5-diphenyltetrazolium bromide (MTT) assay is a widely used colorimetric method for assessing cell metabolic activity, viability, proliferation, and cytotoxicity. It relies on metabolically active cells enzymatically reducing the tetrazolium salt MTT into insoluble purple formazan crystals (Fig. 1). While early studies suggested that succinate dehydrogenase, a mitochondrial enzyme, played a central role in MTT reduction, it is now well-established that this process involves multiple enzymatic systems. These include enzymes located in the endoplasmic reticulum, NAD(P)H-dependent enzymes, and glycolytic enzymes [2–5].

Viable cells, which contain active metabolic enzymes, convert MTT into formazan, while nonviable cells do not. The formed formazan crystals are dissolved using a solvent such as dimethyl sulfoxide (DMSO), producing a colored solution. The intensity of the color, measured at 570 nm using a spectrophotometer, is directly proportional to the number of viable cells [4, 5].

Paula Alvito and Helena Oliveira (eds.), *Food Contaminants Toxicity*, Methods and Protocols in Food Science,
https://doi.org/10.1007/978-1-0716-5096-7_9,

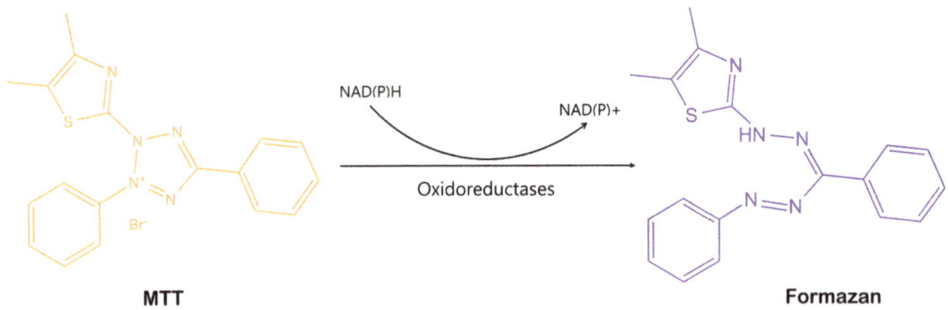

Fig. 1 Representation of the MTT reduction reaction. MTT (yellow) is reduced into purple formazan crystals (purple). (Adapted from [3])

Although widely used, the MTT assay, like other methods, presents limitations that must be considered (*see* **Note 1**). Since this colorimetric assay relies on the enzymatic reduction of MTT by cellular dehydrogenases, its output reflects metabolic activity rather than actual cell viability, number, or proliferation [6]. Therefore, while the MTT assay remains a valuable tool, its suitability must be carefully assessed according to experimental objectives and the specific cell models employed.

Despite these considerations, the MTT assay remains a fundamental technique in cell biology due to its simplicity, reliability, and cost-effectiveness. It is widely used to calculate the half-maximal inhibitory concentration (IC_{50}), which measures the potency of a substance in inhibiting a biological process by 50%. Accurate IC_{50} determination enhances the assay's utility in evaluating the efficacy and safety of new compounds.

Beyond its widespread use in pharmacology and toxicology, the MTT assay also plays a crucial role in assessing the toxicity of food contaminants. It provides quantitative data on cell viability to help determine critical concentrations and establish safety thresholds. Substances such as mycotoxins, pesticides, and heavy metals are examples of chemicals recognized by the World Health Organization (WHO) as major public health concerns [7–9]. Studies have shown that mixtures of mycotoxins and pesticides can exhibit stronger toxicity than individual compounds, highlighting the importance of assays like MTT in understanding these interactions [10, 11]. Similarly, lead, cadmium, arsenic, or methylmercury toxicity has also been evaluated using the MTT assay [12–15]. Advances in MTT assay sensitivity will further enhance its application in food safety and risk assessment. Overall, the MTT assay offers a rapid and quantitative method for assessing cell viability and metabolic activity. It provides crucial data for understanding cellular responses under various conditions and supports advances in drug development and toxicological research.

Optimizing key parameters, such as cell plating and density, solution preparation, washing techniques, replicates, and incubation times for different cell types and experimental conditions (e.g., temperature, CO_2 levels), is essential to ensure reproducibility and reliability [4].

2 Materials

2.1 Reagents

1. 3-(4,5-dimethylthiazol-2-yl)-2,5-diphenyltetrazolium bromide (MTT, 298-93-1).
2. Phosphate-buffered saline (PBS).
3. Dimethyl sulfoxide (DMSO, 67-68-5).
4. Culture medium (e.g., DMEM).
5. Sodium dodecyl sulfate (SDS, 151-21-3).
6. Testing compound.

2.2 Equipment

1. 96 well flat-bottomed microplate tissue-culture grade.
2. Microplate reader with 570 nm filter.
3. Incubator (37 °C, 5% CO_2).
4. 100 μL, 200 μL, and 1000 μL pipettes.
5. Multichannel pipettes (optional).
6. Sterile pipette tips.
7. Sterile cell culture flasks.
8. Cell counter.
9. Orbital shaker.
10. Sterile 20 μm syringe filter or similar.
11. Class II biological safety cabinet (minimum).

3 Methods

3.1 Cell Culture

1. Plate cells in a 96 well plate at the appropriate density (typically 1×10^4 to 5×10^4 cells per well, depending on the growth rate of the cell) and incubate overnight to allow adhesion (*see* **Notes 2–6** and Fig. 2 for an example of a plate setup).

3.2 Cell Treatment

1. Remove the medium with caution to not disturb the cells (*see* **Notes 7 - Removing solutions** and 7 - **Minimizing cell disruption**).
2. Treat cells with (*see* Fig. 2 and **Notes 4** and **6**):

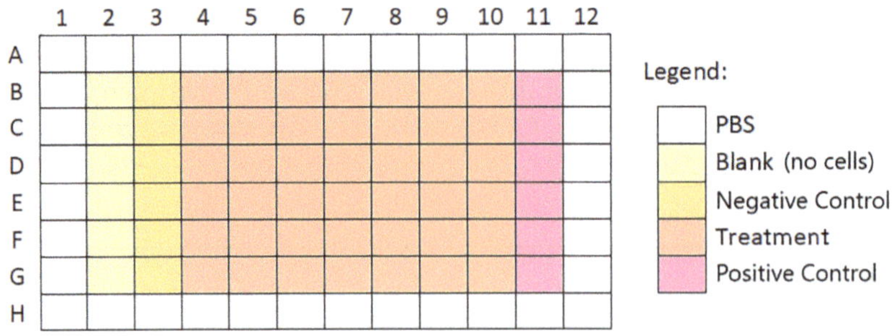

Fig. 2 Example of a 96-well plate setup for the MTT assay. Experimental setup showing the distribution of controls and treatment conditions. Blank wells (without cells) used to subtract background absorbance. Negative control wells contain untreated cells, positive control wells contain cells treated with a known cytotoxic agent, and treatment wells contain cells exposed to different concentrations of the test compound

- Experimental compounds.
- Negative control (untreated cells or solvent control, as appropriate).

3. Incubate for the desired period (typically 24, 48 and 72 h)
4. One hour before the end of the incubation period, treat the wells designated as positive control with 0.1% (v/v) SDS in culture medium for 1 hour (see Figure 2 and **Note 4 - Positive control**). After this treatment, proceed immediately to the MTT incubation step.

3.3 MTT Solution Preparation (See Note 8)

1. Dissolve MTT in PBS to prepare a 5 mg/mL stock solution.
2. Filter the stock solution with a 20 μm syringe filter or similar.
3. Dilute the stock solution in the culture medium to a final concentration of 0.5 mg/mL.

3.4 MTT Incubation

1. Remove the cell supernatant after the incubation period and wash the wells twice with PBS (*see* **Note** 7).
2. Add 100 μL of 0.5 mg/mL MTT solution to each well and incubate for 2–4 h at 37 °C (*see* **Note 9**)
3. Under a microscope, check the wells for purple formazan crystals (*see* Fig. 3)

3.5 Formazan Solubilization and Measurement

1. Carefully remove the MTT solution without disturbing the formazan crystals and wash the wells twice with PBS (*see* **Note** 7).
2. Add 100 μL of DMSO to each well to dissolve the formazan crystals.
3. Incubate under agitation for 10–20 min at 37 °C protected from light.

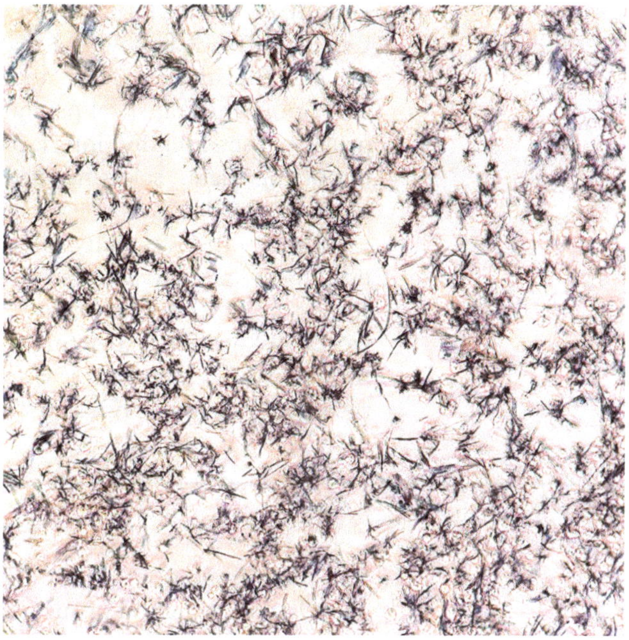

Fig. 3 Formation of purple formazan crystals in SH-SY5Y cells observed under a phase-contrast microscope (Leica model DMIL HC) after incubation with MTT solution. The image highlights metabolically active cells capable of reducing MTT into formazan crystals. Magnification: 200×

4. Before reading, gently shake the plate to ensure complete solubilization and homogeneity.
5. Measure the absorbance at 570 nm using a microplate reader.

3.6 Data Treatment

1. Cell Viability Calculation

   Cell viability is calculated as the percentage of living cells in treated samples relative to the untreated control. This is typically determined using the following formula:

$$Cellviability(\%) = \frac{\text{Abs}_{\text{treatment}} - \overline{\text{Abs}_{\text{blank}}}}{\overline{\text{Abs}_{\text{negativecontrol}}} - \overline{\text{Abs}_{\text{blank}}}} \times 100$$

2. Calculation of IC_{50}

   The IC_{50} value is a critical parameter that indicates the concentration of a compound required to inhibit cell viability by 50%. Although several approaches can be used to determine the IC_{50}, the choice of method depends on the quality of the experimental data and the tools available to the researcher.

   To ensure accurate and reproducible results, we recommend using software designed explicitly for nonlinear regression and dose-response analysis. Commonly used tools include

GraphPad Prism [16] and online calculators such as the IC_{50} Calculator provided by AAT Bioquest [17], which allows users to fit logistic models and extract IC_{50} values with corresponding statistical parameters. These platforms typically provide user-friendly interfaces and guided workflows for importing data, selecting fitting models, and interpreting results.

Due to the variability of available software and interfaces, no step-by-step procedure is described in this chapter. Instead, readers are encouraged to use the platform that best fits their workflow and data structure. The use of reliable tools with robust fitting algorithms is essential to obtain meaningful and biologically relevant IC_{50} values (*see* **Notes 10** and **11**).

4 Notes

1. The main limitations of the MTT assay are as follows: mitochondrial dysfunction or metabolic alterations can affect MTT reduction independently of actual cell death, leading to misleading results. Elevated levels of reactive oxygen species, particularly superoxide, may artificially enhance formazan production, resulting in an overestimation of cell viability [18]. The assay cannot distinguish between cytostatic and cytotoxic effects, as non-dividing but metabolically active cells can still reduce MTT [19, 20]. Compared to assays that measure ATP content, membrane integrity, or DNA, MTT may provide less reliable results when cellular metabolism does not accurately reflect cell viability or proliferation [21].
2. Depending on the cell type, it may be necessary to pre-coat the plates.
3. To reduce poor reproducibility and low signal and maintain consistent metabolic activity across wells, achieve uniform cell seeding and optimal cell density:
   - Uniform Cell Seeding: Ensure cells are evenly dispersed by gently resuspending the cell suspension before transferring it to the plate. If available, use a multi-channel pipette to minimize variation during plating.
   - Optimal Cell Density: Verify that cell density falls within the optimal range for the assay. Overcrowding can restrict the proper diffusion of reagents and nutrients, while insufficient density may result in inadequate metabolic activity and weak signals.
   - Room Temperature Pre-incubation (optional): For certain cell types [CHO-hIR IRS1-GFP [22] and SH-SY5Y [23]], it was shown that a room temperature pre-incubation of the

96-well plate in a Class II biological safety cabinet for 1 h before transferring it to the incubator improved the uniformity of cell distribution within wells by minimizing aggregation around the well edges and enhancing plating homogeneity. Uneven distribution, often caused by immediate movement of the plate, can affect cell behavior, complicate analysis, and reduce reproducibility.

4. Blank and Control Wells (*see* Fig. 2 for an example of a plate setup):
   - Blank: Include blank wells (e.g., wells without cells and containing medium) to measure and subtract background absorbance from the final measurements. This step ensures that the absorbance readings accurately reflect the metabolic activity of the cells.
   - Negative control: Wells containing cells exposed to culture medium alone (untreated control) or to culture medium supplemented with the same concentration of solvent used to prepare the test compound (solvent control). This control ensures that any cytotoxicity observed is attributed to the test compound and not to the solvent itself. The negative control establishes the baseline metabolic activity corresponding to 100% cell viability, allowing for comparison with the treated groups.
   - Positive control: Apply a known cytotoxic substance, such as 0.1% SDS (the concentration may need to be adjusted depending on the cell type), one hour before the end of the treatment phase. SDS induces rapid and complete cell death and should not be incubated for extended periods, serving as a control for maximal cytotoxicity.
5. Use at least four wells per condition, including blanks and both positive and negative controls. Perform all tests in multiple replicates to reduce variability and improve statistical reliability. *See* Fig. 2 for an example of a plate setup.
6. Add PBS to the outer wells to minimize evaporation effects.
7. Procedure for Removing Residual Medium and Solutions:
   - Removing solutions: After incubation, carefully invert the 96-well plate to decant the medium or MTT solution. Use a pipette with gentle suction to remove any remaining liquid from each well to avoid disrupting the cell layer. This step minimizes cross-contamination and ensures uniformity across wells. Note: In polarized cells, be extra cautious with this process.
   - Washing with PBS: A thorough washing step is essential to minimize poor reproducibility and high background absorbance. Add an appropriate volume (e.g., 100 μL) of PBS to

each well to eliminate residual compounds. Invert the plate again and aspirate the PBS thoroughly with a pipette. Repeat the washing process if necessary to ensure complete removal of residuals.

- It is recommended not to use culture media with phenol red, since phenol red can interfere with absorbance readings in the MTT assay by contributing to background absorbance at the measurement wavelength (570 nm). This can result in inaccurate readings and affect the assay's sensitivity. Therefore, phenol red-free media should be used. If the culture medium contains phenol red, extra caution should be taken during the washing steps to remove residual phenol red from the wells.
- Minimizing Cell Disruption: Avoid direct contact with the bottom of the well while aspirating liquid to prevent disturbing adherent cells. This precaution is particularly crucial for fragile or loosely adherent cell types.

8. MTT Solution Storage and Handling:
   - Fresh Preparation: The MTT working solution should be freshly prepared to ensure accuracy and minimize low signal.
   - Stock Solution Storage: The stock solution can be stored at 2–8 °C for less than 2 weeks and at −20 °C for several months (always confirm storage conditions with the supplier for optimal stability).
   - Light Protection: MTT reagent (powder) and the solutions (stock and working) must be protected from light to maintain stability and prevent degradation.
   - Complete Dissolution: Always ensure the MTT is fully dissolved before use to avoid variability in results.
9. Incubate the MTT solution for 2–4 h, depending on the cell type and the experimental conditions. The incubation time should be optimized based on the cell's metabolic activity and the expected formation of formazan crystals. Typically, 2 h is sufficient for most cell lines, but some may require more prolonged incubation time for optimal signal detection.
10. When using online IC_{50} calculators or specialized software (e.g., AAT Bioquest, GraphPad Prism), it is important to review the output critically. Most tools offer goodness-of-fit indicators, such as R^2 values or residual plots, and confidence intervals, which should be considered when validating the results. Additionally, users should be aware of the selected fitting model (e.g., four-parameter logistic model) and ensure that it aligns with the biological behavior of the experimental data.

11. In biological assays, negative viability or response values are typically not meaningful. Therefore, it is often appropriate to restrict regression models to the positive response domain. One way to do this is by fixing the minimum response to zero, effectively simplifying the four-parameter logistic (4PL) model into a three-parameter logistic (3PL) model. This approach improves biological interpretability and avoids artefactual fitting of negative values.

Acknowledgments

The author, Paula Alvito, acknowledges the financial support of national funds through FCT/MCTES under the project CESAM - Centro de Estudos do Ambiente e do Mar, references UID/50017/2025 (doi.org/10.54499/UID/50017/2025) and LA/P/0094/2020 (doi.org/10.54499/LA/P/0094/2020).

References

1. Mosmann T (1983) Rapid colorimetric assay for cellular growth and survival: application to proliferation and cytotoxicity assays. J Immunol Methods 65:55–63. https://doi.org/10.1016/0022-1759(83)90303-4
2. Berridge MV, Herst PM, Tan AS (2005) Tetrazolium dyes as tools in cell biology: new insights into their cellular reduction. Biotechnol Annu Rev 11:127–152. https://doi.org/10.1016/S1387-2656(05)11004-7
3. Khalef L, Lydia R, Filicia K, Moussa B (2024) Cell viability and cytotoxicity assays: biochemical elements and cellular compartments. Cell Biochem Funct 42. https://doi.org/10.1002/cbf.4007
4. Ghasemi M, Turnbull T, Sebastian S, Kempson I (2021) The mtt assay: utility, limitations, pitfalls, and interpretation in bulk and single-cell analysis. Int J Mol Sci 22. https://doi.org/10.3390/ijms222312827
5. Stockert JC, Horobin RW, Colombo LL, Blázquez-Castro A (2018) Tetrazolium salts and formazan products in cell biology: viability assessment, fluorescence imaging, and labeling perspectives. Acta Histochem 120:159–167. https://doi.org/10.1016/j.acthis.2018.02.005
6. Rai Y, Pathak R, Kumari N, Sah DK, Pandey S, Kalra N, Soni R, Dwarakanath BS, Bhatt AN (2018) Mitochondrial biogenesis and metabolic hyperactivation limits the application of MTT assay in the estimation of radiation induced growth inhibition. Sci Rep 8:1–15. https://doi.org/10.1038/s41598-018-19930-w
7. Exposure to dioxins and dioxin-like substances: a major public health concern. https://www.who.int/publications/i/item/WHO-CED-PHE-EPE-19.4.4. Accessed 7 May 2025
8. Public health impact of pesticides used in agriculture. https://iris.who.int/handle/10665/39772. Accessed 7 May 2025
9. Mycotoxins. https://www.who.int/news-room/fact-sheets/detail/mycotoxins. Accessed 7 May 2025
10. Skrzydlewski P, Twarużek M, Grajewski J (2022) Cytotoxicity of mycotoxins and their combinations on different cell lines: a review. Toxins (Basel) 14. https://doi.org/10.3390/toxins14040244
11. Alehashem M, Peters R, Fajana HO, Eslamizad S, Hogan N, Hecker M, Siciliano SD (2024) Herbicides and pesticides synergistically interact at low concentrations in complex mixtures. Chemosphere 353:141431. https://doi.org/10.1016/j.chemosphere.2024.141431
12. Li Z, Lei Z, Wang Y, Wang C, Xu J, Wang Z (2014) Studying cytotoxicity of low concentration arsenic on PC 12 cell line. Anal Methods 6:1709–1713. https://doi.org/10.1039/c3ay42086h
13. Wang H, Chen B, He M, Yu X, Hu B (2017) Selenocystine against methyl mercury

cytotoxicity in HepG2 cells. Sci Rep 7:1–8. https://doi.org/10.1038/S41598-017-00231-7
14. Aziz R, Rafiq MT, Yang J, Liu D, Lu L, He Z, Daud MK, Li T, Yang X (2014) Impact assessment of cadmium toxicity and its bioavailability in human cell lines (Caco-2 and HL-7702). Biomed Res Int 2014:839538. https://doi.org/10.1155/2014/839538
15. Li N, Cui N, Qiao M, Shen Y, Cheng Y, Song L, Huang X, Li L (2022) The effects of lead exposure on the expression of IGF1R, IGFBP3, Aβ40, and Aβ42 in PC12 cells. J Trace Elem Med Biol 69. https://doi.org/10.1016/j.jtemb.2021.126877
16. Prism – GraphPad. https://www.graphpad.com/features. Accessed 13 May 2025
17. AAT Bioquest, Inc. Quest GraphTM IC50 Calculator. https://www.aatbio.com/tools/ic50-calculator. Accessed 27 Jan 2025
18. Wang S, Yu H, Wickliffe JK (2011) Limitation of the MTT and XTT assays for measuring cell viability due to superoxide formation induced by nano-scale TiO2. Toxicol In Vitro 25: 2147–2151. https://doi.org/10.1016/J.TIV.2011.07.007
19. Ghasemi M, Turnbull T, Sebastian S, Kempson I (2021) The mtt assay: utility, limitations, pitfalls, and interpretation in bulk and single-cell analysis. Int J Mol Sci 22:12827. https://doi.org/10.3390/IJMS222312827/S1
20. Riss TL, Moravec RA, Niles AL, Duellman S, Benink HA, Worzella TJ, Minor L (2016) Cell viability assays. In: Assay Guidance Manual. https://www.ncbi.nlm.nih.gov/books/NBK144065/. Accessed 9 Jul 2025
21. Wang P, Henning SM, Heber D (2010) Limitations of MTT and MTS-based assays for measurement of antiproliferative activity of green tea polyphenols. PLoS One 5:e10202. https://doi.org/10.1371/JOURNAL.PONE.0010202
22. Lundholt BK, Scudder KM, Pagliaro L (2003) Technical notes: a simple technique for reducing edge effect in cell-based assays. SLAS Discovery 8:566–570. https://doi.org/10.1177/1087057103256465
23. Dravid A, Raos B, Svirskis D, O'Carroll SJ (2021) Optimised techniques for high-throughput screening of differentiated SH-SY5Y cells and application for neurite outgrowth assays. Sci Rep 11:1–15. https://doi.org/10.1038/s41598-021-03442-1

Chapter 10

Advanced In Vitro Models to Assess Contaminants Toxicity by Ingestion

Fernanda Rosário and Helena Oliveira

Abstract

The use of three-dimensional (3D) models, such as spheroids, offers a valuable approach to mimicking the physiological and pathological characteristics of human gastric tissues. These models are particularly relevant for studying food contamination and ingestion-related toxicity. In this protocol, Spheroids are formed in nonadherent agarose-coated 96-well plates by seeding defined cell densities, without cell adherence to the substrate, thus promoting self-aggregation. Over a 7-day culture period, spheroid diameter is monitored daily to assess compaction and stability. This method results in uniform spheroids ranging between 150 and 350 μm, suitable for downstream toxicological assays. This protocol is cost-effective, user-friendly, and adaptable to various experimental conditions.

Key words Spheroids, 3D models, Food contamination, Ingestion models, Gastric toxicity

1 Introduction

Ensuring food safety is a major public health concern, as exposure to chemical contaminants through ingestion can lead to severe health risks. Many foodborne contaminants, including heavy metals, mycotoxins, and pesticide residues, can accumulate in human tissues and disrupt metabolic processes [1, 2]. To evaluate their toxicity, reliable in vitro models are needed to replicate the physiological conditions of the human digestive system. Traditionally, in vivo animal models have been widely used for toxicity testing. However, these methods are costly, time-consuming, and raise ethical concerns regarding the use of laboratory animals. The European Union has established strict regulations on animal testing through Directive 2010/63/EU, encouraging the development of alternative in vitro models that reduce or replace animal experimentation [3]. In vitro toxicity studies have historically relied on two-dimensional (2D) monolayer cell cultures, where cells adhere to a flat plastic surface. While 2D models offer simplicity and

Paula Alvito and Helena Oliveira (eds.), *Food Contaminants Toxicity*, Methods and Protocols in Food Science,
https://doi.org/10.1007/978-1-0716-5096-7_10,

reproducibility, they fail to accurately reflect the complex microenvironment of human tissues. Cells grown in 2D exhibit morphological alterations, reduced cell-to-cell interactions, and altered signaling pathways [4]. Furthermore, they lack the three-dimensional (3D) structure necessary for the development of physiological diffusion gradients and tissue-specific architecture [5]. As a result, 2D cultures often provide limited predictive value for in vivo toxicity outcomes. 3D cell culture models, such as spheroids and organoids, have emerged as a more biologically relevant alternative [6, 7]. These models promote improved cell-cell and cell-extracellular matrix interactions, allowing for better replication of in vivo conditions [8, 9]. Compared to 2D cultures, 3D spheroids demonstrate enhanced cellular differentiation, prolonged viability, and metabolic activity that more closely resemble those of native tissues [4, 5]. This makes them particularly useful for long-term studies and repeated-dose toxicity assessments. Among the available gastric models, the AGS (human gastric adenocarcinoma) cell line is widely used due to its robustness and adaptability in culture. AGS cells are commonly employed in studies of gastric cancer biology, drug screening, and toxicity assessments [10, 11]. Despite being a cancer-derived cell line, AGS cells retain key characteristics of gastric epithelial cells, making them a suitable model for investigating the effects of ingested toxicants on the gastric mucosa [12]. In this chapter, we present a detailed protocol for generating AGS-derived spheroids as a reliable in vitro model for assessing the toxicity of food contaminants. The method outlined here ensures reproducibility, cost-effectiveness, and ease of implementation. By employing a 3D culture system, this approach enables a more accurate prediction of cytotoxic, genotoxic, and metabolic responses, improving risk assessment strategies in food safety.

2 Materials

2.1 Cell Culture and Sub-culture Materials

All liquids and plasticware used for cell culture must be sterile. In this protocol, the human gastric adenocarcinoma cell line AGS (ATCC) was used and cultured in a humidified atmosphere at 37 °C with 5% CO_2. During the culture period, cell confluence and morphology were monitored using an inverted phase-contrast microscope (Nikon Eclipse TS100, Japan). When confluence reached approximately 80%, cells were subcultured using 0.25% trypsin/1 mM EDTA.

1. 75 cm^2 flasks for cell culture.
2. Serological pipettes.
3. 15 mL conical centrifuge tubes.
4. Hemocytometer.

5. Trypan Blue solution 0.4%, sterile-filtered.
6. 10 and 1000 μL micropipette and pipette tips.
7. Complete growth medium for AGs cells: Ham's F-12 K (Kaighn's) low glucose growth medium supplemented with 10% (v/v) fetal bovine serum (FBS), 1% of L-glutamine, 1% of penicillin-streptomycin).
8. Phosphate-buffered saline (PBS, 1×) pH 7.4, sterile-filtered: 155 mM NaCl, 1.54 mM KH_2PO_4, 2.7 mM Na_2HPO_4-$7H_2O$.
9. Trypsin– EDTA: 0.25% (trypsin–1 mM) EDTA without Ca^{2+} & Mg^{2+} in Hank's balanced salt solution.
10. Inverted microscope.
11. Incubator (37 °C, 5% CO_2).
12. Centrifuge.
13. Class II biological safety cabinet.

2.2 Spheroid Formation Materials

1. Schott flask.
2. Autoclave.
3. Flat bottomed 96 well plates.
4. Multichannel pipet (50–300 μL).
5. Multichannel reagent boat.
6. 100 μL autoclavable pipette tips
7. Deionized water (dH_2O).
8. 1.5% (w/v) agarose solution: The agarose solution is prepared by dissolving 0.15 g of low melting point agarose in 10 mL of deionized water (dH_2O) in a Schott flask. Seal with aluminum foil or lid and autoclave for 20 min at 121 °C. Store at 4 °C until use. (*see* **Note 1**).

3 Methods

3.1 AGS Cell Culture Preparation

1. Culture AGS cells in complete growth medium in 75 cm^2 flasks under standard conditions (37 °C, 5% CO_2), replacing the medium every 48 h (*see* **Note 2**).
2. Once cells reach 80–90% confluence, discard the medium and gently wash the monolayer with pre-warmed PBS to remove residual serum.
3. Add room-temperature Trypsin-EDTA (0.25%) and incubate the flask for 1–3 min at 37 °C.
4. Observe cells under a microscope. When detachment is evident, add two volumes of pre-warmed complete growth

medium or Trypsin Neutralizing Solution to inactivate the trypsin (*see* **Note 2**).

5. Transfer the cell suspension to a 15 mL centrifuge tube.
6. Centrifuge at 200× *g* for 3 min and discard the supernatant.
7. Resuspend the pellet in 5 mL of complete culture medium and collect 10 μL for cell counting.
8. Mix 100 μL of the cell suspension with an equal volume of 0.4% Trypan Blue solution.
9. Load the hemocytometer chamber and examine under the microscope.
10. Count viable cells (unstained) and nonviable cells (blue-stained) in the four large corner squares.
11. Calculate the cell density using the formula: $((A + B + C + D)/4) \times 10{,}000 = \text{cells/mL}$.
12. Adjust the suspension to a final concentration of 4000 cells per well using complete medium.

3.2 96-Well Plates Coating

1. Melt the agarose: Take the pre-prepared 1.5% (w/v) low-melting-point agarose solution from 4 °C storage (*see* **Note 3**). To melt, place the container, loosely covered with a lid, in a microwave and heat at 600–800 W in short intervals of 15–30 s. Gently swirl between each interval. Continue until the solution is fully liquefied and clear. Do not allow the solution to boil, as this may lead to evaporation and alter the final concentration.
2. To prevent premature solidification during dispensing, keep the melted agarose at approximately 60 °C throughout the coating process. This can be achieved by placing the flask in a sterile water bath or heating block, covered with aluminum foil to maintain sterility.
3. Under sterile conditions, pour melted agarose in a multichannel reagent boat.
4. Using a multichannel pipette, quickly dispense 50 μL of melted agarose into each well of a flat-bottom 96-well plate (*see* **Note 4**).
5. Allow the plates with agarose cool down to room temperature before use (*see* **Note 5**).

3.3 Spheroid Generation

1. On day 1 seed 100 μL of cell suspension (4000 cells per well) into the pre-agarose coated plates (*see* **Note 6**). (See Fig. 1 for spheroid size vs. seeding density over time).
2. Incubate the plates under standard culture conditions (37 °C, 5% CO_2).

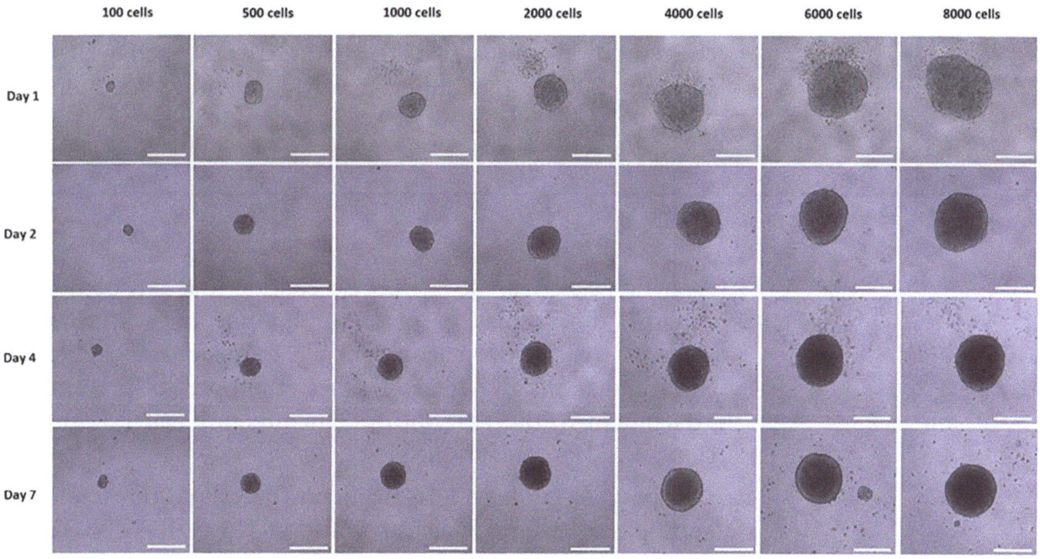

Fig. 1 Optimization of spheroid density

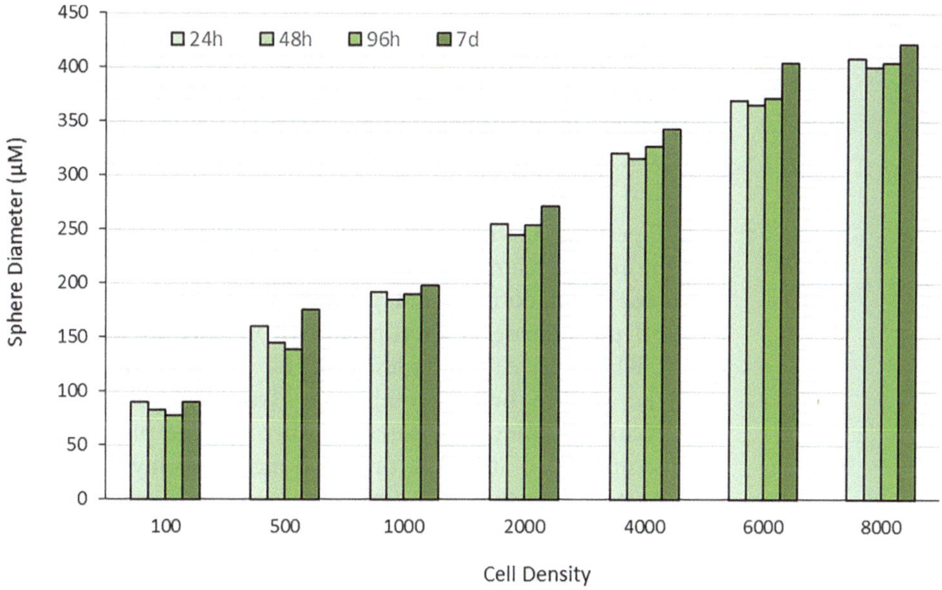

Fig. 2 Spheroid diameter variation with incubation time

3. At 24 h post-seeding, observe spheroids under an inverted microscope to assess early aggregation (see Fig. 2 for diameter monitoring).
4. At 48 h, compact spheroids should have formed. Replace 50% of the medium with fresh complete growth medium (*see* **Notes 7** and **8**).

5. At 72 h, prepare treatment solutions at 2× the desired final concentration and add 100 μL directly to each well (final volume: 200 μL), without removing existing medium (*see* **Note 9**).

4 Notes

1. Agarose role: Agarose is a polysaccharide that can be used to support the formation of the spheroids once it forms a non-cytotoxic hydrogel, mimicking as much as possible the extracellular matrix. Its non-adhesive properties prevent cell attachment to the well bottom, encouraging cell–cell interaction and 3D aggregation.
2. Cell passage guidance: To maintain the cell cultures, add the appropriate volume of cell suspension to new 75 cm^2 flasks. As a general guide, from a confluent flask of cells: 1:2 split should be 70–80% confluent and ready for an experiment in 1–2 days. 1:5 split should be 70–80% confluent and ready for an experiment in 2–4 days.
3. Agarose storage: The 1.5% (w/v) agarose solution can be stored at 4 °C for up to 10 days and reheated before use as described in Subheading 3.2, **Step 1**. Troubleshooting: If agarose gels unevenly or too quickly, verify reheating procedure and pipetting speed.
4. Pipetting agarose: The agarose solidifies within seconds to minutes, ensuring pipette tips are used swiftly to prevent agarose from cooling and gelling inside the tips. A volume of 50 μL agarose is appropriate to produce a concave surface. CRITICAL Amounts less than 50 μL may not entirely cover the surface before solidifying. Troubleshooting: If agarose coating is uneven or patchy, ensure tips and plates are at room temperature and pipetting is done smoothly. Tip: For even coating, gently tap the plate on a sterile surface immediately after pipetting.
5. Storage of the Plates: Plates can be used immediately or stored up to about 10 days after preparation if hermetically sealed (e.g., with parafilm), at room temperature and protected from light. Tip: Before use, inspect plates for dryness or contamination and discard if compromised.
6. Cell seeding and compactness: An important consideration for spheroid culture is the choice of optimal cell numbers. The goal is to use the minimum number of cells necessary to produce stable aggregates of a sufficient size for easy handling and analysis and not generation necrotic core. Cell density will vary depending on cell line and doubling time. To estimate

the minimal cell density for producing viable aggregates with stable shape and sufficient size, spheroids with cell numbers ranging from 100 to 8000 per single spheroid were prepared and culture them for 7 days (Fig. 1). Troubleshooting: If spheroids are not compact after 48 h, check that agarose coating is uniform and that cell suspension was evenly distributed; If spheroids are too small or fragmented, increase cell density or improve cell viability; If spheroids adhere to the plate edges, ensure agarose fully covers the well bottom and surface is intact.

7. Size guidelines: To ensure spheroid quality and reproducibility, both morphological criteria and size thresholds should be considered. To avoid the formation of a necrotic core, spheroids should ideally reach diameters between 150 and 350 μm after 48 h of incubation. The optimal spheroid size depends on the experimental objective: For drug testing or 3D cell–cell and cell–matrix interaction studies, small spheroids (<200 μm) may be sufficient. To replicate physiological conditions such as hypoxia and proliferation gradients without central necrosis, spheroids of 350–400 μm (typically reached by 96 h) are recommended. Larger spheroids (500–600 μm) often develop central necrosis, which may be useful in certain studies but undesirable in others. To monitor spheroid development, diameter was measured over 7 days using images acquired by bright-field microscopy with a calibrated scale bar (Fig. 2). Measurements were performed using ImageJ software. After 48 h, spheroids usually become more compact compared to 24 h, due to natural cellular compaction driven by cell–cell adhesion and cytoskeletal reorganization. Troubleshooting: If spheroids exceed 400 μm and necrosis is not intended, reduce seeding density or incubation time; If spheroids remain small, check cell health and ensure accurate counting and seeding.

8. Spheroid quality assessment parameters: Diameter: 150–350 μm after 48 h, with a coefficient of variation (CV) <15% across replicate wells. Morphology: Round and symmetrical spheroids with smooth, well-defined borders. Compactness: Dense internal structure, without visible cavities or fragmentation. Circularity: Preferably >0.85, calculated using ImageJ. These parameters help define "mature" spheroids and provide a robust foundation for assessing reproducibility across experiments. Troubleshooting: Irregular shapes or low circularity may indicate uneven seeding or agarose coating defects.

9. Reproducibility: To confirm uniformity, measure spheroid diameters across ≥6 replicate wells. A coefficient of variation (CV) <15% was considered acceptable for intra-experimental uniformity. Troubleshooting: Higher variability suggests

inconsistent plate preparation, uneven seeding, or incubation issues; review these steps carefully.

Acknowledgments

This work is funded by national funds through FCT—Fundação para a Ciência e a Tecnologia I.P., under the project/grant UID/50006 + LA/P/0094/2020 (doi.org/10.54499/LA/P/0094/2020). FCT is acknowledged for the research contract under Scientific Employment Stimulus to F.R. (2022.04220.CEECIND/CP1720/CT0043; DOI:10.54499/2022.04220.CEECIND/CP1720/CT0043) and H.O. (CEECIND/04050/2017/CP1459/CT0023; DOI:10.54499/CEECIND/04050/2017/CP1459/CT0023).

References

1. Persistent Organic Pollutants in Food: Contamination Sources, Health Effects and Detection Methods – PMC (n.d.) https://pmc.ncbi.nlm.nih.gov/articles/PMC6888492/. Accessed 3 Apr 2025
2. The Sources of Chemical Contaminants in Food and Their Health Implications – PMC (n.d.) https://pmc.ncbi.nlm.nih.gov/articles/PMC5699236/. Accessed 3 Apr 2025
3. Directive – 2010/63 – EN – EUR-Lex (n.d.) https://eur-lex.europa.eu/legal-content/EN/TXT/?uri=celex:32010L0063. Accessed 3 Apr 2025
4. Duval K, Grover H, Han LH, Mou Y, Pegoraro AF, Fredberg J, Chen Z (2017) Modeling physiological events in 2D vs. 3D cell culture. Physiology 32:266–277. https://doi.org/10.1152/PHYSIOL.00036.2016
5. Imamura Y, Mukohara T, Shimono Y, Funakoshi Y, Chayahara N, Toyoda M, Kiyota N, Takao S, Kono S, Nakatsura T, Minami H (2015) Comparison of 2D- and 3D-culture models as drug-testing platforms in breast cancer. Oncol Rep 33:1837–1843. https://doi.org/10.3892/or.2015.3767
6. Ravi M, Paramesh V, Kaviya SR, Anuradha E, Paul Solomon FD (2015) 3D cell culture systems: advantages and applications. J Cell Physiol 230:16–26. https://doi.org/10.1002/JCP.24683
7. Huang X, Huang Z, Gao W, Gao W, He R, Li Y, Crawford R, Zhou Y, Xiao L, Xiao Y (2022) Current advances in 3D dynamic cell culture systems. Gels. https://doi.org/10.3390/gels8120829
8. Baker EL, Srivastava J, Yu D, Bonnecaze RT, Zaman MH (2011) Cancer cell migration: integrated roles of matrix mechanics and transforming potential. PLoS One 6:e20355. https://doi.org/10.1371/JOURNAL.PONE.0020355
9. Boudreau N, Bissell MJ (1998) Extracellular matrix signaling: integration of form and function in normal and malignant cells. Curr Opin Cell Biol 10:640–646. https://doi.org/10.1016/S0955-0674(98)80040-9
10. Manente L, Sellitti A, Lucariello A, Laforgia V, de Falco M, de Luca A (2011) Effects of 4-nonylphenol on proliferation of AGS gastric cells. Cell Prolif 44:477–485. https://doi.org/10.1111/J.1365-2184.2011.00774.X
11. Hall AJ, Tripp M, Howell T, Darland G, Bland JS, Babish JG (2006) Gastric mucosal cell model for estimating relative gastrointestinal toxicity of non-steroidal anti-inflammatory drugs. Prostaglandins Leukot Essent Fatty Acids 75:9–17. https://doi.org/10.1016/J.PLEFA.2006.04.006
12. Savino L, Di Marcantonio MC, Moscatello C, Cotellese R, Centurione L, Muraro R, Aceto GM, Mincione G (2022) Effects of H2O2 treatment combined with PI3K inhibitor and MEK inhibitor in AGS cells: oxidative stress outcomes in a model of gastric cancer. Front Oncol 12:860760. https://doi.org/10.3389/FONC.2022.860760/FULL

Chapter 11

Integrated In Vitro Genotoxicity Testing Protocols for Safety Assessment of Food Nanomaterials After Simulated Human Digestion

Henriqueta Louro, Nádia Vital, Maria Cardoso, Paula Alvito, and Maria João Silva

Abstract

Nanomaterials (NMs), such as titanium dioxide (TiO_2), are already widely used, and many innovative types, such as nanocelluloses (CNMs), are under development for applications in food and feed industry, requiring a thorough safety assessment regarding ingestion. This chapter describes the protocol for an integrated approach that uses a static in vitro digestion protocol to mimic human digestion prior to a battery of genotoxicity tests adapted to NMs, covering mammalian gene mutations, numerical, and structural chromosome aberrations. The inclusion of the digestion simulation in the design of the safety testing strategy can be seen as a new approach methodology (NAM) contributing to bridge the in vitro to in vivo physiology, providing important data for hazard assessment without the use of animal models.

Key words Nanomaterials, New approach methodology, In vitro digestion, Nanotoxicology, Genotoxicity, Micronucleus, Gene mutation

1 Introduction

Nanotechnology is a transformative technology with the potential to enhance a wide range of products and processes across various fields, including medicine, agriculture, and the food and feed industry. Numerous nanomaterials (NMs), such as titanium dioxide (TiO_2), are already in use, and many more innovative types, like cellulose nanomaterials (CNMs), are currently under development for the food and feed sector [1]. However, for product development, the safety assessment of nanomaterials involves specific challenges related to their small size, surface reactivity, and consequent dynamic behavior under biological conditions [2]. International bodies and regulatory agencies have thrived in the last years to define adequate procedures for the risk assessment (RA) of NMs [3, 4]. In view of the relevant concern with the potential of NMs to

Paula Alvito and Helena Oliveira (eds.), *Food Contaminants Toxicity*, Methods and Protocols in Food Science,
https://doi.org/10.1007/978-1-0716-5096-7_11,

affect the carcinogenic process, their genotoxic effects have been brought to research highlights. Recent work has focused on considerations for ingested NMs using conventional methodologies, introducing modifications to account for nano-specific properties (reviewed in [1, 5]). The latest developments in Europe suggest the use of in vitro simulated digestion protocols to adequately consider the dissolution/degradation rate of NMs in conditions representative of the human gastrointestinal tract (GIT), as a key first step in the stepwise approach for RA of NMs to be applied in food or feed [6]. Several protocols have been described for in vitro digestion simulation, using either static or dynamic settings. By incorporating a step of static in vitro simulated digestion before the toxicological testing of NMs, the integrated protocol described in this chapter intends to proxy the physiological modifications that NMs suffer in the GIT after oral intake. The obtained digested NMs can be used both for physicochemical characterization and for toxicity testing in human intestinal cell models, as depicted in Fig. 1. In addition, 2D or 3D gastrointestinal cell models can be used to further mimic human GIT, providing a promising new approach methodology (NAM) to replace animal models.

The NMs dispersion should be based on standardized protocols (*see* **Note 1**). Due to the tendency of NMs to form large agglomerates or aggregates when in aqueous media, and to bind to protein in the cell media, it is generally recommended that the

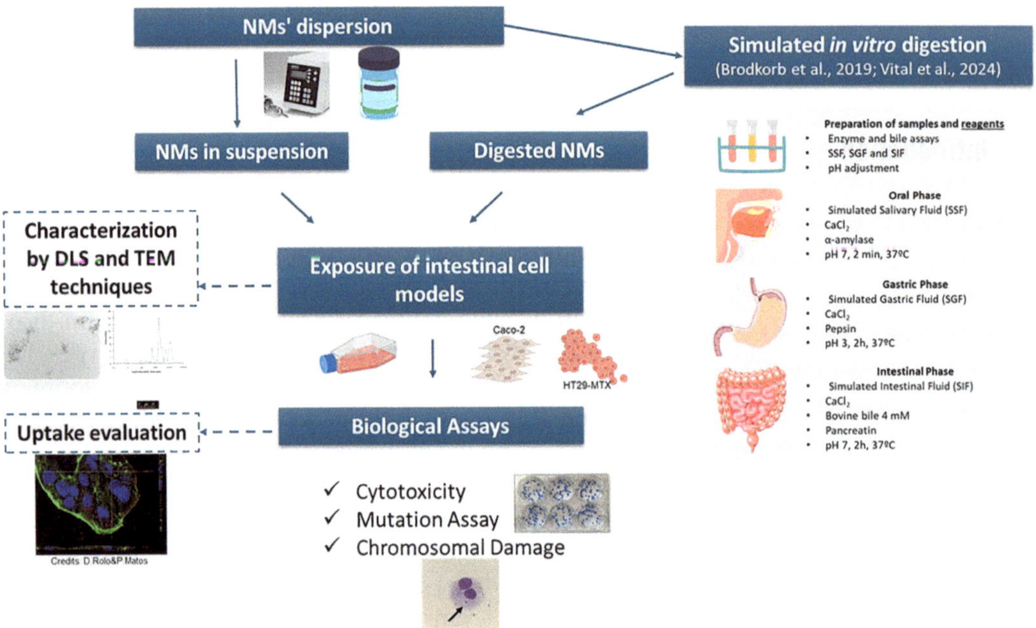

Fig. 1 Flow chart illustrating the integrated experimental design for studying the effects of nanomaterials (NMs) on intestinal cell models, assembling the INFOGEST 2.0 in vitro digestion. In vitro digestion protocol, modified for intestinal fluid composition. DLS dynamic light scattering, TEM transmission electron microscopy

quality of the dispersion of NMs for biological testing should be carefully confirmed. Several methods can be used for characterizing the NMs' dispersions to warrant the nano-scale properties, such as dynamic light scattering (DLS) or electron microscopy (EM) [7].

Dispersed NMs are directly added to cells in culture or submitted to the in vitro digestion, with the resulting digested product used to expose the cells. The INFOGEST 2.0 in vitro digestion model is a standardized protocol developed by the COST INFOGEST network; it is used to simulate human digestion, with parameters based on available physiological data [8, 9]. Briefly, the protocol simulates the oral, gastric, and intestinal phase of digestion, and, for each phase, a specific simulated digestion fluid is prepared—simulated salivary fluid (SSF), simulated gastric fluid (SGF), and simulated intestinal fluid (SIF), respectively. Each fluid is composed of corresponding electrolyte stock solutions, enzymes, bile salts (only for SIF), $CaCl_2$, and water. This method has been applied for mycotoxin digestion in food [10] and then adapted for NMs in collaborative projects at the National Institute of Health Dr. Ricardo Jorge, Lisboa, Portugal [11, 12], and recently was modified [13] to the version described in this protocol. Therefore, a modified SIF with a bile salt concentration of 4 mM may be used instead of the standard 10 mM concentration specified in the INFOGEST 2.0 digestion protocol [13]. Remaining steps of the original protocol are as described in Brodkorb et al. [8].

The characterization of the changes produced in the NMs following in vitro digestion is an important step, and simple methods, such as DLS, or more complex methodologies, such as scanning electron microscopy (SEM), transmission electron microscopy (TEM), or single particle inductively coupled plasma mass spectrometry (sp-ICP-MS), can be used (*see* **Note 2**).

With the knowledge of the NMs' properties in the digested product, namely solubility/dissolution rate and particle size distribution, decisions can be made whether to use conventional genotoxicity tests or use methodologies modified to account for the presence of NMs (EFSA Scientific Committee, 2021). The latter situation means that the toxicological assays must take on board nanoscale considerations, as recommended by OECD [3]. The battery of tests to evaluate NM's genotoxicity should provide information on three endpoints: gene mutation, structural, and numerical chromosome damage [14, 15].

Concerning gene mutation, the Ames' test is widely used; however, it is considered as producing false-negative results for NMs, due to the lack of capacity to cross the bacterial wall according to OECD [3]. Thus, a mammalian cell gene mutation assay is recommended and was included in the present protocol. The endogenous hypoxanthine-guanine phosphoribosiltransferase (*Hprt*) gene is a chromosome X-linked gene that codifies for the protein HPRT, an enzyme that is involved in the purine salvage pathway, where it recycles guanine to guanosine monophosphate during DNA degradation. In the gene mutation assay, the *Hprt* gene is used as a target and reporter gene allowing determination of the frequency and spectrum of mutations induced by the test substance. The addition of a selective agent, 6-thioguanine (6-TG), allows the selection of mutant cells. It is an analog of purines that, when taken up by cells, is metabolized through the salvage recycling pathway, catalyzed by HPRT, to form a toxic metabolite that leads to cell death. Point mutations (e.g., base pair substitutions, frameshifts, small deletions, and insertions) and deletions of the gene impairing the gene function block the salvage recycling pathway of purines and allow mutant cells to survive in the selective medium. When a positive result is observed, it is possible to recover the mutant colonies and analyze the mutation spectrum by sequencing the gene, as described elsewhere [16]. The method described in this section is based on TG 476 [17] and has been applied to investigate the mutagenicity of NMs by the authors in recent research projects [18, 19].

For detecting both structural and numerical chromosome damages, the micronucleus assay is usually applied according to TG487 [20]. The micronucleus assay is based on the formation of small intracytoplasmic nuclei containing whole chromosomes that detached from the spindle apparatus (aneugenesis) or chromosome fragments (clastogenesis) upon cell division by mitosis. The addition of cytochalasin B to the cell culture during the exponential growing phase of cells blocks cytokinesis and, consequently, in the end of cell division the two newly formed nuclei share the same cytoplasm. This ensures that micronuclei scored in so-called binucleated cells were formed during the precedent mitosis, presumably due to the genotoxic effect of the test substance to which cells were exposed. Guidance on adaptations necessary for testing NMs includes delaying the time for addition of the cytokinesis-blocking agent (cytochalasin B) that may inhibit NM uptake, focusing on long exposure, no need to include extracellular metabolic activation system (S9 mix), using detection by flow cytometry with care as optically active nanomaterials can interfere with the scoring, and considering visual interference under manual scoring [3, 6, 21]. The method described in this chapter is based on OECD TG 487 [20] and in recent guidance [21] and has been applied to NMs by the authors in recent research projects, linked to human digestion [12, 18].

Notably, it is important to select the appropriate cell lines for testing genotoxicity, since the uptake is an important feature that will drive cellular effects (*see* **Note 3**). Intracellular uptake of the NMs should be demonstrated, and only cells that can efficiently internalize NMs should be used for micronucleus test [21].

This chapter describes an integrative approach that assembles two genotoxicity tests, covering the necessary endpoints for investigating genotoxicity of the NMs in 2D intestinal cell models, after the use of the static INFOGEST 2.0 in vitro digestion protocol as adapted to NMs. The genotoxicity tests described are the *Hprt* gene mutation and the micronucleus assays. The appropriate dose range to use for testing the genotoxicity of NMs is based on preliminary dose-range-finding assays (*see* **Note 4**).

The approach here has the important advantage of allowing a comprehensive genotoxicity assessment at early stages of NMs development, and the combination of the in vitro digestion simulation in the design of the safety testing strategy can be seen as a NAM, improving nanosafety assessment accounting for the digestion process, without the need of animal studies in compliance with the 3Rs principles [22].

2 Materials

2.1 Cell Culture and Exposure to the Test NM

1. Test Nanomaterial (NM).
2. Probe Sonicator: 400 Watt Sonifier equipped with a standard 13 mm disruptor horn.
3. Intestinal cell line, such as Caco-2, HT29-MTX-E12, or Chinese hamster lung V79 cells (required for HPRT gene mutation assay)—authenticated cell lines, free of mycoplasma contamination.
4. Trypsin-EDTA 0.05%.
5. Dulbecco's Phosphate-Buffered Saline (DPBS).
6. Dimethylsulfoxide (DMSO).
7. Complete cell culture medium, depending on cell type. For Caco-2 or HT29-MTX- E12 cells: Dulbecco's Modified Eagle Medium (DMEM) with 1% penicillin/streptomycin (10,000 U penicillin/10 μg streptomycin) supplemented with 10% Fetal Bovine Serum (FBS) and 2.5% of 4-(2-hydroxyethyl)-1-piperazineethanesulfonic acid (HEPES) buffer. For V79 Cells: DMEM or MEM α supplemented with 1% penicillin/streptomycin (10,000 U penicillin/10 μg streptomycin), 10% FBS, and 2.5% HEPES buffer.
8. Tissue culture flasks.
9. Standard cell culture material and equipment.

2.2 Gene Mutation Assay

1. Chinese hamster lung V79 cells.
2. Inverted phase contrast microscope.
3. DMEM medium with low glucose, with L-glutamine.
4. MEM α medium + GlutaMAX™.
5. 6-Thioguanine(6-TG) 98%.

6. Sodium carbonate (Na_2CO_3).
7. Fetal Bovine Serum (FBS).
8. HEPES Buffer.
9. Trypan blue solution (0.4%).
10. Methylene blue.
11. HAT supplement: sodium hypoxanthine (5 mM), aminopterin (20 μM), and thymidine (0.8 mM).
12. Cell culture media with HAT: HAT supplement is added to the culture medium at the mutant cleansing stage. Concentrated HAT medium (50× concentrated) may be purchased. Before use, prepare 1× HAT media by diluting 1 volume of HAT Supplement (50×) in 49 volumes of cell culture medium prepared with DMEM.
13. Sodium carbonate solution 0.5% (Na_2CO_3): Prepared by diluting 1 g in 200 mL of tri-distilled H20. Sterilize by autoclaving 5 min, 0.3 atm. Store at room temperature.
14. 6-Thioguanin (6-TG, 500 μg/mL): Dissolve 10 mg of 6-TG in 20 mL of 0.5% of Na_2CO_3, prepared fresh. Sterilize by autoclaving 5 min in 0.3 atm.
15. Methylene blue (1% v/v): Dissolve 1 g of methylene blue in 100 mL of bidistilled water. Keep stock solutions at room temperature. Filter before use. No need to be sterile.
16. Chemical positive control: e.g., Ethylmethanesulfonate (EMS, 3 mg/mL) (*see* **Note 5**).
17. Sterile Petri dishes (100 × 15 mm).
18. Tissue culture 6-well plates and flasks.
19. Neubauer or Bürker Counting Chamber and automated cell counter (optional).

2.3 Micronucleus Assay

1. Light microscope including 40× and 100× objectives.
2. Cytochalasin B (Cyt-B) solution: prepare stock solution at 2 mg/mL in DMSO, aliquot and store at −20 °C.
3. Hypotonic solution (0.1 M KCl): weight 3.73 g of KCl in 500 mL of distilled water.
4. Fix solution: prepare freshly methanol: acetic acid solution (3: 1) (*see* **Note 6**).
5. Giemsa staining solution (4% v/v): prepare 250 mL Giemsa staining solution by adding, at RT, 240 mL of Gurr's phosphate buffer with 10 mL of filtered Giemsa. Protect from light.
6. Gurr's phosphate buffer (pH 6.8): prepare by dissolving Gurr tablets in deionized water.

7. Chemical positive control: e.g., Mitomycin C (MMC, 1 mg/mL) (*see* **Notes** 7 and **8**).
8. Entellan mounting medium.
9. Tissue culture 6-well plates and flasks.
10. Microscope Glass slides, Glass Coverslips.

3 Methods

3.1 *In Vitro* Mammalian Cell Gene Mutation Test Using the Hprt Gene

1. *Hprt spontaneous mutant cleansing.*

   Prior to the experiment, to ensure that spontaneous Hprt– mutant cells are eliminated, thaw a vial of V79 cells and sub-culture 2–4 times in HAT medium. After cleansing, cells can be proceeded to the assay or can be frozen in aliquots to be used for further mutation experiments.
2. *Day 1—Cell seeding before exposure.*
   1. Detach cells from the tissue culture flask: remove the medium, wash twice with 2 mL PBS, add 2 mL trypsin-EDTA, incubate 2 min at 37 °C, check that cells are floating, and then add 4 mL of complete DMEM medium to inactivate trypsin.
   2. Transfer the cell suspension to a 15 mL tube and centrifuge at 1000 rpm, 5 min; remove the supernatant and resuspend the pellet in 4 mL medium.
   3. Determine the cell concentration in the suspension by using a counting chamber or another method available.
   4. Prepare a V79 cell suspension at a density of 1×10^5 cells/mL.
   5. Plate 1×10^6 cells in each 75 cm^2 culture flask, the number of flasks depending on the planned treatment conditions, plus negative and positive controls.
   6. Incubate 24 h at 37 °C, 5% CO_2.
3. *Day 2—Cell exposure.*
   7. At 24 h after seeding check if cells are attached and with a morphology indicating that they are actively dividing. If cells are not dividing, cancel the assay (and repeat day 1).
   8. Prepare the stock dispersion of the NM (*see* **Note 1**). Dilute the dispersion of the NM in complete cell medium, according to the concentrations to be tested (treatment medium). Prepare at least 4 concentrations of NMs for cell exposure. Include also negative, vehicle, and positive controls.
   9. Remove the medium from the flasks, add the treatment medium, and incubate for 24 h at 37 °C, 5% CO_2.

10. The positive control cells are treated on day 3 because they are exposed to EMS (3 mg/mL), for 30 min only (*see* **Note 5**).

4. *Day 3—Plating for the phenotypic expression period and determination of the cloning efficiency (CE).*

11. Recover the cells from the tissue culture flask repeating procedures 1–3.

12. Prepare cell suspensions of each treatment condition at a density of 1×10^5 cells/mL Account for the volume necessary for the plating for procedures 13.

13. *Plating for determination of cloning efficiency in non-selective medium (CE)*: Seed cells on one 6-well plate per treatment condition, at a density of 50 cells per well, and complete with medium up to 2 mL per well; incubate the plates at 37 °C, 5% CO_2, for 5–7 days.

14. *Plating for phenotypic expression*: From the cell suspension at 1×10^5 cells/mL (prepared in 12), distribute 3×10^5 cells per 100 mm Petri dish, using 3 dishes per treatment, complete with medium up to 10 mL, and incubate at 37 ° C, 5% CO_2, for 2–3 days.

5. *Days 6/7—First Sub-culture of cells in phenotypic expression.*

15. Detach cells using trypsin (*see* Subheading 3.1, **step 2(1)**), remove medium after centrifugation and resuspend the pellet in complete medium.

16. Pool together cells from the three dishes per treatment, and determine the cell concentration (*see* Subheading 3.1, **step 2(3)**).

17. Distribute 3×10^5 cells per 100 mm Petri dish, using three dishes per treatment, complete up to 10 mL with culture medium, and incubate at 37 °C, 5% CO_2, for 2–3 days.

6. *Days 8/9—Second sub-culture of cells in phenotypic expression.*

18. Repeat **steps 15–17** to maintain cells in exponential growth during the period of phenotypic expression.

7. *Days 11/13—Plating for the first mutant frequency (MF1) determination in selective medium and first CE (CE1) determination in nonselective medium.*

19. At the end of the phenotypic expression period (8–10 days), recover the cells from the Petri dishes subcultured in **step 18**.

20. Pool together the three dishes per treatment, suspend, and determine the cell concentration per treatment.

21. For each treatment condition, prepare cell suspensions at a density of 1×10^5 cells/mL in complete MEM-α medium

to start the selection period and, in parallel, the CE determination.

22. *Plating for determination of MF1*: From the cell suspensions prepared in 21, distribute approximately 3×10^5 cells per 100 mm Petri dish, using 6 dishes per treatment, add 10 mL of MEM-α medium to each dish, and incubate at 37 °C, 5% CO_2, for 2–3 h until cells start to attach.
23. Add the selective agent 6-TG (5 μg/mL), and incubate for 8–10 days, at 37 °C, 5% CO_2.
24. *Plating for CE1*: In parallel, seed 50 cells per well of a 6-well plate (1 plate per treatment condition), add 2 mL complete MEM-α medium (without 6-TG) to each well and incubate at 37 °C, 5% CO_2, for 5–7 days.
25. Distribute 3×10^5 cells of the remaining cell suspensions prepared in 21 per 100 mm Petri dish, using 3 dishes per treatment, add 10 mL of MEM-α medium to each dish, and incubate at 37 °C, 5% CO_2, for 2 days. These cells will be used to determine MF2 and CE2.

8. *Days 13/15—Plating for the second MF (MF2) determination in selective medium and second CE (CE2) determination in nonselective medium:*

26. Recover cells from the Petri dishes cultured in **step 25** and repeat **steps 21–25**.

9. *Staining.*

27. At the end of the appropriate time for CE (5–7 days after plating, in nonselective medium) and MF determination (8–10 days after plating in selective medium), stain the colonies by adding 1% methylene blue directly to the culture media in each well of 6-well plates or each 100 mm Petri dish, and allow to stain for 20 or 30 min, respectively.
28. Rinse with water twice, invert the plates and let them dry at room temperature.
29. Count the number of nonoverlapping colonies composed by at least 50 cells in each well of 6-well plates or Petri dish, and register. Count the number of colonies formed in 100 mm Petri dish (colonies of mutant cells).

10. *Data analysis.*

30. For the calculations, the following equations can be used:

Cloning efficiency (CE):

$$CE(\%) = \frac{(\text{number of colonies})}{(\text{number of plated cells})} \times 100$$

The cytotoxicity of the treatment is measured by the relative survival, or relative cloning efficiency (RCE) immediately after treatment, and is calculated according to the equation:

$$RCE(\%) = \frac{(CE \text{ of exposed cultures})}{(CE \text{ of unexposed cultures})} \times 100$$

Mutant Frequency (MF):

$$MF = \frac{\text{number of mutant colonies}}{(\text{number of plated cells} \times CE \text{ in nonselective medium})}$$

The MF is adjusted for the number of plated cells that are able to form colonies by using the CE determined at the beginning of the selection period. Two selection periods are described in the protocol, being the first one the standard procedure and the second one to allow the recovery of cells harboring *Hprt* mutations that need a longer period for their expression. The results of both harvests are pooled together to determine the MF. Replicates from 2 or 3 independent experiments are used to determine the mean and standard error for each exposure condition.

31. The criteria for the acceptability of the results are as recommended by the OECD TG 476 [17] and include appropriate comparison with positive and negative controls, and with the historical control data from the same laboratory (*see* **Notes 9** and **10**).

3.2 *In Vitro* Micronucleus Assay for Testing NMs

1. *Day 1—Preparation of the cell culture in exponential growth phase.*
   1. Detach the intestinal cells from the tissue culture flask: remove the medium, wash with 2 mL Trypsin-EDTA, add 2 mL trypsin-EDTA, incubate 3–4 min at 37 °C, check that cells are floating, and then add 6 mL of complete medium.
   2. Dilute cells 1:3 and complete up to 10 mL with medium.
   3. Incubate overnight at 37 °C, 5% CO_2.
2. *Day 2—Cell seeding before exposure.*
   4. Recover the cells from the tissue culture flask repeating the procedure in 1.
   5. Transfer to a 15 mL tube and centrifuge at 1000 rpm, 5 min, remove the supernatant, resuspend the pellet in 4 mL medium.
   6. Determine the cell concentration in the suspension using a counting chamber or another method available.
   7. Prepare cell suspensions at a density of 1×10^5 cells/mL.
   8. Plate cells in 6-well plates or in 25 cm^2 flasks according to planned treatment conditions (*see* **Notes 11** and **12**).
   9. Incubate 24 h at 37 °C, 5% CO_2.

3. *Day 3—Cell treatment and incubation.*
   10. After the attachment period, about 24 h after seeding, check the cell growth, which should be on exponential phase.
   11. Prepare the stock dispersion of the NM.
   12. Dilute the dispersion of the NM in cell culture medium, according to the concentrations to be tested.
   13. Remove the medium from the flasks/wells and add the treatment medium and incubate 24 h at 37 °C, 5% CO_2.
4. *Day 4—Cytokinesis block with cytochalasin B (Cyt B).*
   14. At 23 h, prepare MMC work solution (WS) at 10 μg/mL. Add MMC to the positive control cell culture according to the type of cell used (*see* **Note 8**).
   15. At 24 h after start of exposure, add Cyt B to each flask/well, without discarding the medium, to obtain a final concentration of 4.5 μg/mL.
   16. Incubate for additional 1.5–2 cell cycles (e.g., 28 h for Caco-2) at 37 °C, 5% CO_2.
5. *Day 5—End of treatment, harvesting of cells, slide preparation, fixation, and staining.*
   17. At 52 h, i.e., 28 h after Cyt B addition, rinse the cells twice with 2 mL PBS, add fresh medium with 10% FBS to each well/flask, and incubate for 1 h at 37 °C.
   18. Recover the cells from the wells, suspend the cells by gently pipetting, and transfer the total volume of each culture to a 15 mL centrifuge tube.
   19. Centrifuge at 800 rpm for 5 min. Discard the supernatant by inversion of the tube. Suspend the pellet by gentle finger-tapping.
   20. Add 1 mL of the hypotonic solution, at room temperature, drop-by-drop, while vortexing softly the tube.
   21. Immediately add 2 mL of cold fixing solution, drop-by-drop while gently vortexing.
   22. Centrifuge the tubes at 800 rpm for 5 min.
   23. Discard the supernatant and resuspend the pellet by gentle finger-tapping.
   24. Add 5 mL of cold fixing solution drop-by-drop while gently vortexing.
   25. Incubate at room temperature for 10 min.
   26. Centrifuge 5 min at 800 rpm. Discard the supernatant by inversion of the tube and leave about 300 μL of liquid.

27. Suspend the cells and spread them by gently projection with the pipette tip horizontally touching the glass microscope slide.
28. Leave the slides to dry at room temperature for 1–2 days.
29. After slides are air-dried, wash by immersion in Gurr's phosphate buffer (pH 6.8) and stain by immersion in Giemsa at 4% in phosphate buffer, for 5–15 min (*see* **Note 13**).
30. Finally, wash slides with Gurr's phosphate buffer, twice, allow them to air-dry overnight.
31. Use 2–3 drops of Entellan mounting medium to fix the slide covers.

6. *Slide analysis and results.*

32. Slides are analyzed under a bright field microscope after coding all slides to allow blind scoring and avoid observer bias. Example photographs are shown in Fig. 2.
33. The cytokinesis-block proliferation index (CBPI) is determined by counting the number of mono-, bi-, and multinucleated cells in at least 500 cells per culture (40× magnification), to assess cell proliferation (OECD 2023).

$$\mathrm{CBPI} = \frac{(\mathit{No}.\text{mononucleated cells} + 2 \times \mathit{No}.\text{binucleated cells} + 3 \times \mathit{No}.\text{multinucleated cells})}{(\text{Total number of cells scored})}$$

34. The micronucleus is scored at 400× or 1000× magnification, in at least 2000 binucleated cells per concentration (1000 binucleated cells per culture; two cultures per concentration).

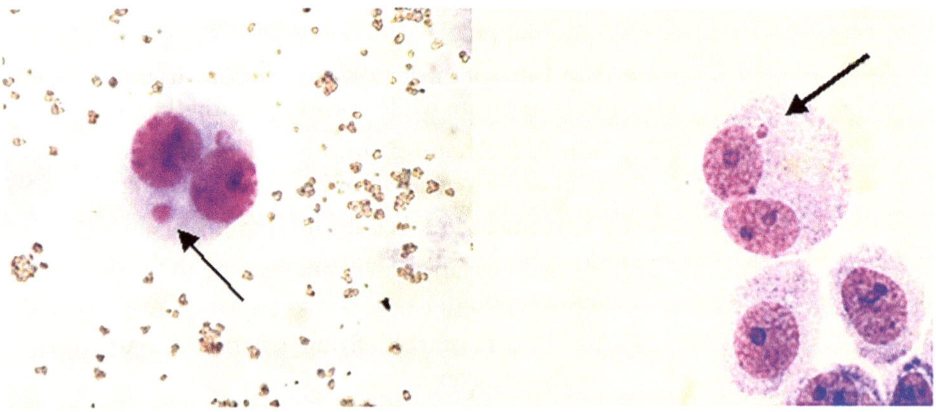

Fig. 2 Example of microscopy image of slides for scoring MN. (**a**) Binucleated cell exposed to TiO_2NPs, showing one micronucleus and one nucleoplasmic bud; (**b**) Cells exposed to nanocellulose, showing mononucleated cells and one binucleated cell with one micronucleus (magnification 400×)

35. General considerations on scoring:
    - Cells containing more than two main nuclei should not be analyzed for micronuclei, as the baseline micronucleus frequency may be higher in those cells.
    - Scoring of mononucleate cells is acceptable if the test substance is shown to interfere with Cyt B activity.
    - The full description of scoring criteria can be found at [20].
36. Appropriate statistical analysis of data should be performed, e.g., Fisher's exact test for the micronucleus frequency or One-Way Anova for CBPI data. Regression analysis may be used to determine a dose-response relationship.
37. The criteria for the acceptability of the results are as recommended by the OECD TG 487 [20] and include appropriate comparison with positive and negative controls, and with the historical control data from the same laboratory (*see* **Notes 9** and **10**).

4 Notes

1. *Nanomaterials preparation*: The standard NANOGENOTOX protocol has been adopted under international bodies [23] and is based on probe sonication at defined energy to warrant appropriate homogeneous dispersion of the insoluble NMs after pre-wetting the powder in 0.5% absolute ethanol (96%), followed by the addition of sterile-filtered 0.05 wt% of bovine serum albumin (BSA)-water. It allows reasonable dispersion of the test materials immediately after its production and can ensure stable dispersion up to 1 h allowing to perform exposure of cells before onset of agglomeration and/or sedimentation. In case of fiber-like NMs, particularly for aqueous suspensions of CNMs, there is no consensus on the adequate methodological dispersion procedure to be used, as sonication may lead to the break of long fibers, thus not reflecting the real scenario [23]. Dispersion by applying vortex and characterization after dispersion in ultrapure water, PBS, or directly on cell culture are a minimal requirement [23–26].
2. *Characterization of the NMs*: For measuring the diameter and length of fibers, SEM and TEM are currently the only methods advised [7]. The recommended methods have been incorporated into OECD testing guideline 125 [7] and detailed guidance can also be found, e.g., at EFSA website [4, 27, 28]. The most common approach to test the solubility

involves (ultra)filtration and guidance for the testing of solubility, dissolution rate, and dispersion stability of NMs is available [29].

3. *Cellular uptake of NMs*: The in vitro genotoxicity testing of NMs should always include an assessment of cellular uptake, especially to substantiate negative test results [6]. Methodologies for assessment of the internalization of NMs by cells, including TEM and confocal microscopy, include SEM or TEM with energy dispersive x-ray (EDX) detection [30, 31], or with the support of confocal microscopy, if fluorescent NMs are available [32]. The cellular uptake evaluation of gold, silica, and silver nanoparticles in five different cell lines, including intestinal cells, has been published by the Joint Research Centre [33].
4. *Preliminary cytotoxicity testing*: A preliminary assay for determining appropriate concentrations to test should be undertaken. Current guidelines indicate that, if no cytotoxicity is observed, NMs should be tested up to 100 μg/mL or 100 μg/cm^2, whichever is highest [21]. Higher concentrations may imply agglomeration of the NMs that will change size distribution and also may lead to precipitation. In addition, higher concentrations are not physiologically relevant, and the micronucleus assay can result in interference with scoring due to high deposition on cells [21]. Cytotoxicity can be tested by different assays, such as the plating efficiency (PE) or the cytokinesis-block proliferation index (CBPI), which are part of each of the protocols described in the above sections. Alternative protocols can be based on membrane integrity and use membrane impermeable dyes (e.g., Trypan blue exclusion test) or staining with the fluorescent dyes (e.g., propidium iodide) as have been reviewed recently [34]. The description of these protocols can be found in the literature. In case of cytotoxicity, when performing the micronucleus assay, the concentration range should cover noncytotoxic concentrations and the highest concentration should achieve 55 ± 5% cell death [20]. Conversely, if the maximum concentration is based on cytotoxicity, the highest concentration in the HPRT gene assay should aim to achieve between 20% and 10% of relative survival [17], as described in the assay protocol.
5. *Positive controls for HPRT gene mutation assay*: Ethylmethanesulfonate or other chemicals control can be used as positive control, known to induce mutations in this assay, such as Ethylnitrosourea or 4-Nitroquinoline 1-oxide [17].
6. *Solution of fixative* based on methanol and acetic acid should be prepared in a ventilated fume hood with appropriate safety precautions.

7. *Positive controls for micronucleus assay*: Presently, no suitable particle positive controls have been identified. Thus, in the absence of particle positive controls, it is recommended that chemical positive controls continue to be used [20]. Clastogens active without metabolic activation include Methyl methane-sulphonate, Mitomycin C, 4-Nitroquinoline-NOxide, and Cytosine arabinoside. Aneugens recommended are Colchicine and Vinblastine [20].
8. *Concentrations of MMC suggested for micronucleus assay as follows: For Caco-2 cells*, use final concentration of 0.15 μg/mL. Remove the medium from wells and add MMC treatment medium to the respective flask. For HT29-MTX-E12: use final concentration of 0.3 μg/mL. Remove the medium from flasks and add the MMC treatment medium to the respective well.
9. *Acceptability criteria*: The criteria for the acceptability of the results are as recommend by the OECD TG 476 [17] or TG 487 [20] and include appropriate comparison with positive and negative controls, and with the historical control data from the same laboratory.
10. *Interpretation of genotoxicity data*: Following the confirmation of the validation of each assay, the OCDE criteria to identify a positive/negative result should be considered [3, 20], as follows. A response is considered a clear positive in a specific test if it meets all the criteria below in at least one experimental condition:
    - at least one of the data points exhibits a statistically significant increase compared to the concurrent negative control,
    - the increase is concentration- or dose-related at least at one sampling time when evaluated with an appropriate trend test,
    - the result is outside the distribution of the historical negative control data (e.g. Poisson-based 95% control limits).

    According to OECD, the test chemical is considered clearly negative if, in all experimental conditions examined, none of the above criteria for a positive result are met. These criteria apply to both micronucleus and gene mutation assay.

    In cases when the response is neither clearly negative nor clearly positive as described above, or to assist in establishing the biological relevance of a result, the data should be evaluated by expert judgment and/or further investigations.
11. *Number of test concentrations for micronucleus*: According to OECD TG 487, in the micronucleus test at least three concentrations of test item should be tested alongside with negative and positive controls, all in duplicate cultures [20].

12. *Cell suspension concentration and type of culture*: For Caco-2 cells, add 8 mL of the cell suspension to a T25 flask (25 cm^2), to have a final density of approximately 32,000 cells/cm^2; for HT29-MTX-E12, add 2 mL of the cell suspension to a 6-well plate, to have a final density of approximately 21,000 cells/cm^2.
13. *The duration of Giemsa* staining in the micronucleus assay should be tested first in a few slides and decided after checking that the staining allows clear distinction of cytoplasm and nuclei.

Acknowledgments

This research was co-funded by the Portuguese Foundation for Science and Technology, through national funds: UID/04923/2020 – Comprehensive Health Research Centre, PhD grant number 2020.07168.BD (https://doi.org/10.54499/2020.07168.BD). P. Alvito acknowledges FCT/MCTES through national funds CESAM UID/50017/2025 (https://doi.org/10.54499/UID/50017/2025) and LA/P/0094/2020 (https://doi.org/10.54499/LA/P/0094/2020). The authors thank all the support from the colleagues Ricardo Assunção, Carla Martins, and Mafalda Silva (Department of Food and Nutrition, INSA, Lisbon, Portugal) concerning the simulated digestion and from Maria Dušinská (NILU—Norwegian Institute for Air Research) for support with HPRT assay. Dora Rolo and Paulo Matos (INSA, Lisbon, Portugal) are acknowledged for the uptake image included in Fig. 1.

References

1. Vital N, Ventura C, Kranendonk M, Silva MJ, Louro H (2022) Toxicological assessment of cellulose nanomaterials: oral exposure. Nano 12:3375. https://doi.org/10.3390/nano12193375
2. Louro H, Silva MJ (eds) (2022) Nanotoxicology in safety assessment of nanomaterials, advances in experimental medicine and biology, vol 1357, ISBN 978-3-030-88070-5, pp 439. https://doi.org/10.1007/978-3-030-88071-2_10
3. OECD (2018) Evaluation of in vitro methods for human hazard assessment applied in the OECD testing programme for the safety of manufactured nanomaterials, Series on the safety of manufactured nanomaterials No. 85. ENV/JM/MONO(2018)4. OECD Publishing, Paris. https://www.oecd.org/content/dam/oecd/en/publications/reports/2018/03/evaluation-of-in-vitro-methods-for-human-hazard-assessment-applied-in-the-oecd-testing-programme-for-the-safety-of-manufactured-nanomaterials_45226f35/a5da5705-en.pdf
4. EFSA Scientific Committee, More S, Bampidis V, Benford D, Bragard C, Halldorsson T, Hernandez-Jerez A, Bennekou SH, Koutsoumanis K, Lambré C, Machera K, Naegeli H, Nielsen SS, Schlatter J, Schrenk D, Silano V, Turck D, Younes M, Castenmiller J, Chaudhry Q, Cubadda F, Franz R, Gott D, Mast J, Mortensen A, Oomen AG, Weigel S, Barthelemy E, Rincon A, Tarazona J, Schoonjans R (2021) Guidance on technical requirements for regulated food and feed product applications to establish the presence of small

particles including nanoparticles. EFSA J 19(8):6769, 48 pp. https://doi.org/10.2903/j.efsa.2021.6769

5. Vieira A, Gramacho A, Rolo D, Vital N, Silva MJ, Louro H (2022) Cellular and molecular mechanisms of toxicity of ingested titanium dioxide nanomaterials. Adv Exp Med Biol 1357:225–257. https://doi.org/10.1007/978-3-030-88071-2_10
6. EFSA Scientific Committee, More S, Bampidis V, Benford D, Bragard C, Halldorsson T, Herandez-Jerez A, Hougaard Bennekou S, Koutsoumanis K, Lambre C, Machera K, Naegeli H, Nielsen S, Schlatter J, Schrenk D, Silano V, Turck D, Younes M, Castenmiller J, Chaudhry Q, Cubadda F, Franz R, Gott D, Mast J, Mortensen A, Oomen AG, Weigel S, Barthelemy E, Rincon A, Tarazona J, Schoonjans R (2021) Guidance on risk assessment of nanomaterials to be applied in the food and feed chain: human and animal health. EFSA J (19, 8): 6768, 111 pp. https://doi.org/10.2903/j.efsa.2021.6768
7. OECD (2023) Test no. 125: nanomaterial particle size and size distribution of nanomaterials, OECD guidelines for the testing of chemicals, Section 1. OECD Publishing, Paris. https://doi.org/10.1787/af5f9bda-en
8. Brodkorb A, Egger L, Alminger M, Alvito P, Assunção R, Ballance S, Bohn T, Bourlieu-Lacanal C, Boutrou R, Carrière F, Clemente A, Corredig M, Dupont D, Dufour C, Edwards C, Golding M, Karakaya S, Kirkhus B, Le Feunteun S, Lesmes U, Macierzanka A, Mackie AR, Martins C, Marze S, McClements DJ, Ménard O, Minekus M, Portmann R, Santos CN, Souchon I, Singh RP, Vegarud GE, Wickham MSJ, Weitschies W, Recio I (2019) INFOGEST static in vitro simulation of gastrointestinal food digestion. Nat Protoc 14(4):991–1014. https://doi.org/10.1038/s41596-018-0119-1
9. Minekus M, Alminger M, Alvito P, Ballance S, Bohn T, Bourlieu C, Carrière F, Boutrou R, Corredig M, Dupont D, Dufour C, Egger L, Golding M, Karakaya S, Kirkhus B, Le Feunteun S, Lesmes U, Macierzanka A, Mackie A, Marze S, McClements DJ, Ménard O, Recio I, Santos CN, Singh RP, Vegarud GE, Wickham MS, Weitschies W, Brodkorb A (2014) A standardised static in vitro digestion method suitable for food – an international consensus. Food Funct 5(6): 1113–1124. https://doi.org/10.1039/c3fo60702
10. Assunção R, Martins C, Dupont D, Alvito P (2016) Patulin and ochratoxin A co-occurrence and their bioaccessibility in processed cereal-based foods: a contribution for Portuguese children risk assessment. Food Chem Toxicol 96:205–214. https://doi.org/10.1016/j.fct.2016.08.004
11. Bettencourt A, Gonçalves LM, Gramacho AC, Vieira A, Rolo D, Martins C, Assunção R, Alvito P, Silva MJ, Louro H (2020) Analysis of the characteristics and cytotoxicity of titanium dioxide nanomaterials following simulated in vitro digestion. Nanomaterials (Basel) 10(8):1516. https://doi.org/10.3390/nano10081516
12. Vieira A, Vital N, Rolo D, Roque R, Gonçalves LM, Bettencourt A, Silva MJ, Louro H (2022) Investigation of the genotoxicity of digested titanium dioxide nanomaterials in human intestinal cells. Food Chem Toxicol 161:112841. https://doi.org/10.1016/j.fct.2022.11284
13. Vital N, Gramacho AC, Silva M, Cardoso M, Alvito P, Kranendonk M, Silva MJ, Louro H (2024) Challenges of the application of in vitro digestion for nanomaterials safety assessment. Foods 13(11):1690. https://doi.org/10.3390/foods13111690
14. ICH (2013) ICH guideline S2 (R1) on genotoxicity testing and data interpretation for pharmaceuticals intended for human use, EMA/CHMP/ICH/126642/2008. European Medicines Agency, London. Available at: https://www.ema.europa.eu/en/ich-s2-r1-genotoxicity-testing-data-interpretation-pharmaceuticals-intended-human-use-scientific-guideline#related-content-17976
15. EFSA Scientific Committee (2011) Scientific Opinion on genotoxicity testing strategies applicable to food and feed safety assessment. EFSA J 9(9):2379. [69 pp.]. https://doi.org/10.2903/j.efsa.2011.2379
16. Silva MJ, Costa P, Dias A, Valente M, Louro H, Boavida MG (2005) Comparative analysis of the mutagenic activity of oxaliplatin and cisplatin in the Hprt gene of CHO cells. Environ Mol Mutagen 46(2):104–115. https://doi.org/10.1002/em.20138
17. OECD (2016) Test No. 476: In vitro mammalian cell gene mutation tests using the Hprt and xprt genes, OECD guidelines for the testing of chemicals, section 4. OECD Publishing, Paris. https://doi.org/10.1787/9789264264809-en
18. Vital N, Cardoso M, Kranendonk M, Silva MJ, Louro H (2025) Evaluation of the cyto- and genotoxicity of two types of cellulose

nanomaterials using human intestinal cells and in vitro digestion simulation. Arch Toxicol 99(2):575–596. https://doi.org/10.1007/s00204-024-03911-2

19. Vital N, Pinhao M, Yamani NE, Runden-Pran E, Louro H, Dušinska M, Silva MJ (2022) Correction to: hazard assessment of benchmark metal-based nanomaterials through a set of in vitro genotoxicity assays. Adv Exp Med Biol 1357:C1. https://doi.org/10.1007/978-3-030-88071-2_14
20. OECD (2023) Test no. 487: in vitro mammalian cell micronucleus test, OECD guidelines for the testing of chemicals, section 4. OECD Publishing, Paris. https://doi.org/10.1787/9789264264861-en
21. OECD (2022) Study report and preliminary guidance on the adaptation of the in vitro micronucleus assay (OECD TG 487) for testing of manufactured nanomaterials, Series on testing and assessment no. 359, ENV/CBC/MONO(2022)15. OECD Publishing, Paris. https://one.oecd.org/document/ENV/CBC/MONO(2022)15/en/pdf
22. European Commission (2019) Regulation (EU) 2019/1010 of the European Parliament and of the Council of 5 June 2019 on the alignment of reporting obligations in the field of legislation related to the environment, and amending Regulations (EC) No 166/2006 and (EU) No 995/2010 of the European Parliament and of the Council, Directives 2002/49/EC, 2004/35/EC, 2007/2/EC, 2009/147/EC and 2010/63/EU of the European Parliament and of the Council, Council Regulations (EC) No 338/97 and (EC) No 2173/2005, and Council Directive 86/278/EEC (Text with EEA relevance, PE/8/2019/REV/1. Available at http://data.europa.eu/eli/reg/2019/1010/oj
23. Jensen KA, Kembouche Y, Christiansen E, Jacobsen NR, Wallin H, Guiot C, Spalla O, Witschger O (2011) The generic NANOGENOTOX dispersion protocol—standard operation procedure (SOP). https://www.anses.fr/en/system/files/nanogenotox_deliverable_6.pdf
24. Zhang Y, Fox DM, Ong KJ, Ede J, Sayes C, Kangas H, Ilona L, Goergen M, Moon R, Shatkin JA (2024) Limitations of recent cellulose studies: how overzealous dispersion technique, fluorescence labeling and interpretation limit the findings of recent safety assessments for cellulose materials. Trends Food Sci Technol 151:104620. https://doi.org/10.1016/j.tifs.2024.104620
25. Brand W, van Kesteren PCE, Swart E, Oomen AG (2022) Overview of potential adverse health effects of oral exposure to nanocellulose. Nanotoxicology 16(2):217–246. https://doi.org/10.1080/17435390.2022.2069057
26. Pradhan SH, Mulenos MR, Steele LR, Gibb M, Ede JD, Ong KJ, Shatkin JA, Sayes CM (2020) Physical, chemical, and toxicological characterization of fibrillated forms of cellulose using an in vitro gastrointestinal digestion and co-culture model. Toxicol Res (Camb) 9(3):290–301. https://doi.org/10.1093/toxres/tfaa026
27. Aimonen K, Imani M, Hartikainen M, Suhonen S, Vanhala E, Moreno C, Rojas OJ, Norppa H, Catalán J (2022) Surface functionalization and size modulate the formation of reactive oxygen species and genotoxic effects of cellulose nanofibrils. Part Fibre Toxicol 19(1):19. https://doi.org/10.1186/s12989-022-00460-3
28. Bennekou-Hougaard S, Halldorsson TI, Koutsoumanis KP, Allende A, Bearth A, Casacuberta J, Castle L, Coja T, Crèpet A, Hoogenboom L, Knutsen H, Lambré C, Nielsen SS, Turck D, Civera AV, Villa R, Zorn H, Bastos M, Castenmiller J, Cubadda F, Franz R, Gott D, Louro H, Mast J, Mortensen a, Oomen A, Verleysen E, Astuto MC, Cattaneo I, Scattareggia Marchese A, Chaudhry Q (2024a) Annex C—practical instructions for reporting the results of electron microscopy analysis of nanomaterials or materials possibly containing a fraction of small particles including nanoparticles. Annex to: EFSA Scientific Committee, More S, Bampidis V, Benford D, Bragard C, Halldorsson T, Hernandez-Jerez A, Bennekou SH, Koutsoumanis K, Lambré C, Machera K, Naegeli H, Nielsen SS, Schlatter J, Schrenk D, Silano V (deceased), Turck D, Younes M, Castenmiller J, Chaudhry Q, Cubadda F, Franz R, Gott D, Mast J, Mortensen A, Oomen AG, Weigel S, Barthelemy E, Rincon A, Tarazona J and Schoonjans R, 2021. Guidance on technical requirements for regulated food and feed product applications to establish the presence of small particles including nanoparticles – annex C. EFSA J 19(8):6769, 48 pp. https://doi.org/10.2903/j.efsa.2021.6769. Available at: https://efsa.onlinelibrary.wiley.com/action/downloadSupplement?doi=10.2903%2Fj.efsa.2021.6769&file=efs26769-sup-0003_Annex+C_Reporting+EM+Analysis.pdf
29. Bennekou-Hougaard S, Halldorsson TI, Koutsoumanis KP, Allende A, Bearth A, Casacuberta J, Castle L, Coja T, Crèpet A, Hoogenboom L, Knutsen H, Lambré C, Nielsen SS, Turck D, Civera AV, Villa R, Zorn H, Bastos M, Castenmiller J, Cubadda F, Franz R,

Gott D, Louro H, Mast J, Mortensen A, Oomen A, Verleysen E, Astuto MC, Cattaneo I, Scattareggia Marchese A, Chaudhry Q (2024b) Annex D—reporting template for an electron microscopy analysis and results. Annex to: EFSA Scientific Committee, More S, Bampidis V, Benford D, Bragard C, Halldorsson T, Hernandez-Jerez A, Bennekou SH, Koutsoumanis K, Lambré C, Machera K, Naegeli H, Nielsen SS, Schlatter J, Schrenk D, Silano V (deceased), Turck D, Younes M, Castenmiller J, Chaudhry Q, Cubadda F, Franz R, Gott D, Mast J, Mortensen A, Oomen AG, Weigel S, Barthelemy E, Rincon A, Tarazona J and Schoonjans R, 2021. Guidance on technical requirements for regulated food and feed product applications to establish the presence of small particles including nanoparticles. Annex D. EFSA J 2021 19(8):6769, 48 pp. https://doi.org/10.2903/j.efsa.2021.6769. Available at: https://efsa.onlinelibrary.wiley.com/action/downloadSupplement?doi=10.2903%2Fj.efsa.2021.6769&file=efs26769-sup-0004_Annex+D_Template.docx

30. OECD (2020) Guidance document for the testing of dissolution and dispersion stability of nanomaterials, and the use of the data for further environmental testing and assessment, OECD series on testing and assessment, no. 318. OECD Publishing, Paris. https://doi.org/10.1787/f0539ec5-en
31. Louro H, Tavares A, Vital N, Costa PM, Alverca E, Zwart E, de Jong WH, Fessard V, Lavinha J, Silva MJ (2014) Integrated approach to the in vivo genotoxic effects of a titanium dioxide nanomaterial using LacZ plasmid-based transgenic mice. Environ Mol Mutagen 55(6):500–509. https://doi.org/10.1002/em.21864
32. Pinto F, Lourenço AF, Pedrosa JFS, Gonçalves L, Ventura C, Vital N, Bettencourt A, Fernandes SN, da Rosa RR, Godinho MH et al (2022) Analysis of the in vitro toxicity of Nanocelluloses in human lung cells as compared to multi-walled carbon nanotubes. Nanomaterials 12:1432. https://doi.org/10.3390/nano12091432
33. Rolo D, Pereira JFS, Gonçalves L, Bettencourt A, Jordan P, Silva MJ, Matos P, Louro H (2025) Assessing the impact of TiO2 nanomaterials on intestinal cells: new evidence for epithelial translocation and potential pro-inflammatory effects. Toxicology. 511:154066. https://doi.org/10.1016/j.tox.2025.154066
34. Bogni A, Ponti J, Drewes C, Kinsner-Ovaskainen A, Bremer S (2022) In vitro cytotoxicity and cellular uptake evaluation of gold, silica and silver nanoparticles in five different cell lines: Caco-2, A549, CHO, V79 and TK6, EUR 30908 EN. Publications Office of the European Union, Luxembourg, ISBN 978-92-76-44690-3, https://doi.org/10.2760/959826, JRC120791
35. Fenech M, Chang WP, Kirsch-Volders M, Holland N, Bonassi S, Zeiger E, HUman MicronNucleus project (2003) HUMN project: detailed description of the scoring criteria for the cytokinesis-block micronucleus assay using isolated human lymphocyte cultures. Mutat Res 534(1–2):65–75. https://doi.org/10.1016/s1383-5718(02)00249-8

Chapter 12

Adsorption of Potentially Toxic Elements onto Nanomaterials in Aqueous and Biological Media: Supporting Chemical Co-exposure Toxicity Assessment

Ana Teresa Reis, João Paulo Teixeira, and Cláudia Batista Lopes

Abstract

This protocol describes a method for evaluating the adsorption of potentially toxic elements (PTE) onto engineered nanomaterials (ENM). Adsorption studies are performed in ultrapure water and in cell culture media, with or without serum supplementation, to assess both mechanistic interactions and biologically relevant behavior. The protocol includes preparation of ENM dispersions and PTE solutions, adsorption experiments (e.g., isotherm, kinetic, or pH-dependent studies), phase separation, and preparation for quantification analysis. The resulting data provide insight into ENM–PTE interactions and support the interpretation of in vitro co-exposure toxicity assessments. Quantification methods are not covered in this protocol and should be selected according to experimental needs.

Key words Adsorption, Engineered nanomaterials (ENM), Potentially toxic elements (PTE), ENM–PTE interactions, Co-exposure, Acid digestion

1 Introduction

Simultaneous exposure to engineered nanomaterials (ENM) and potentially toxic elements (PTE) through food is an emerging concern [1–5], as these contaminants can interact and produce combined toxicological effects distinct from individual exposures [6–10]. In vitro toxicity studies are commonly used to assess these risks, but proper interpretation requires understanding ENM–PTE interactions, particularly adsorption [6], which can influence solubility, stability, bioavailability, and ultimately, toxicity [11, 12].

This protocol outlines a methodology to assess the adsorption of PTE onto ENM, supporting the interpretation of co-exposure toxicity data. Adsorption studies help distinguish between direct and indirect toxicity by evaluating how physicochemical interac-

Paula Alvito and Helena Oliveira (eds.), *Food Contaminants Toxicity*, Methods and Protocols in Food Science,
https://doi.org/10.1007/978-1-0716-5096-7_12,

tions affect contaminant behavior in biological systems. For example, PTE adsorption may reduce free ion availability or modify nanomaterial surface properties, altering cellular uptake and response [9].

To account for both mechanistic understanding and biological relevance, adsorption studies should be conducted in: (i) ultrapure water—to characterize intrinsic adsorption behavior without interference; and (ii) cell culture media—to reflect realistic experimental conditions and matrix effects such as protein corona formation and competitive binding [13, 14].

This dual-matrix approach is novel in that it integrates mechanistic and biologically relevant environments into a single adsorption workflow, which is not commonly addressed in existing protocols. By combining simplified systems with complex biological media, this protocol fills a critical gap in experimental design for nanotoxicology, enhancing the translational value of in vitro co-exposure data.

The workflow includes: (1) preparation of ENM and PTE solutions; (2) adsorption experiments (e.g., equilibrium, kinetic, isotherm, pH-dependent); (3) phase separation; and (4) quantification of unbound PTE. While the matrix composition differs, the procedure remains similar; digestion may be needed prior to analysis in culture media samples. Quantification of PTE is essential but falls outside the scope of this protocol, as methods vary depending on the specific analyte and instrumentation available.

This dual-approach protocol enables researchers to better interpret in vitro co-exposure results by providing key insights into ENM–PTE interactions under both simplified and biologically relevant conditions [6].

2 Materials

1. All solutions should be prepared using ultrapure water (18 MΩ·cm at 25 °C) and analytical grade reagents. Unless otherwise specified, prepare all solutions and reagents at room temperature.

   All borosilicate and plastic materials must be thoroughly decontaminated prior to use. Acid washing should be carried out by either filling the containers or submerging them in a 25% (v/v) nitric acid solution (prepared by dilution of concentrated nitric acid, HNO_3, 65%) for a minimum of 24 h. Following acid treatment, rinse the materials three times with ultrapure water, and allow them to air-dry in a clean, undisturbed area, preferably under a laminar flow hood or in a designated clean drying cabinet.

Ensure to use all necessary personal protection equipment and to consult Safety Data Sheets to become familiar with the environmental health hazards, protective measures, and safety precautions for handling, storing, and transporting the chemicals. Follow institutional and regulatory guidelines for the proper disposal of chemical waste and contaminated materials.

2.1 Laboratory Materials and Equipment

1. Plasticware and acid-washed borosilicate glassware—centrifuge tubes, pipette tips, Pasteur pipettes, beakers, flasks; volumetric flasks.
2. Analytical balance.
3. Centrifuge (capable of RCF values of 30,000× *g*).
4. Thermoblock (optional).
5. Water bath (optional).
6. Microwave digestion system (optional).
7. Micropipettes and tips.
8. 15 mL centrifuge tubes
9. Borosilicate bottles.
10. Volumetric flasks.
11. 0.22 μm syringe filters
12. Syringes.

2.2 Reagents and Solutions

1. Concentrated nitric acid (HNO_3), 65% analytical grade—for acid-washing, sample acidification, and sample digestion.
2. Ultrapure water (18 MΩ·cm)—for solution preparation, batch experiments, and material rinsing.
3. Selected Potentially Toxic Elements (solution or salt; adapted to the toxicity assessment) (*see* **Note 1**).
4. Selected Engineered Nanomaterials (suspensions or powders; adapted to the toxicity assessment) (*see* **Note 1**).
5. Cell culture media with or without added serum, to match the experimental conditions of toxicity testing.
6. Bovine serum albumin (BSA)—for ENM dispersion procedure: Weigh 0.05 g of BSA in a weighing boat. Transfer the BSA to a 100 mL borosilicate flask, and add 100 mL of ultrapure water. Reserve a few mL of ultrapure water to rinse the weighing boat into the flask to ensure that all BSA was transferred. Gently swirl the flask (do not shake, agitate, or use stir bars, as this will cause foaming and interfere with ENM dispersion stability). Place the flask in the fridge overnight. The next day, again swirl the flask gently and check if all BSA was dissolved. Store at

2–8 °C to prevent microbiological activity (*see* **Note 2**), and avoid stirring or agitation of the solution. Before use, check for possible crystallization or contamination; if confirmed, properly discard the solution and prepare a new one.

7. Ethanol 96%—for ENM dispersion procedure.
8. Ammonium hydroxide (NH_4OH) or sodium hydroxide (NaOH)—for pH adjustment (if needed).
9. Hydrochloric acid (HCl), 37% (m/v) analytical grade—for pH adjustment (if needed).
10. Certified reference materials (optional)—for method validation or recovery studies.

2.3 Engineered Nanomaterials

2.3.1 ENM Dispersion (2.56 mg/mL)

1. In a lidded vial, carefully weigh 15.36 mg of the nanomaterial (*see* **Note 3**) using a cleaned steel or glass spatula. Immediately close the vial and clean the weighing and working area for potential spills using wet and dry wipes.
2. Open the vial with the pre-weighed nanomaterial and tilt it approximately 45°. Using a micropipette, add 30 μL of 96% ethanol drop-by-drop onto the powder. Screw the lid back on and gently mix the ethanol and powder by tapping the vial gently on the tabletop and rotating it from side to side between your fingers for about one minute.
3. Slowly add 5970 μL of 0.05% Bovine Serum Albumin (BSA) solution by pipette, while gently rotating and swirling the tilted vial. Reserve about 2 mL of BSA solution to rinse any nanomaterial powder from the inner wall of the vial. Avoid shaking or agitating the vial to prevent foaming of the BSA.
4. To set up the sonication apparatus (Fig. 1), fill a 500 mL glass beaker with ice and place it upside-down in a 1000 mL beaker. Place a small crystallizing dish on top of the inner beaker. Carefully place the vial (with lid) in the middle of the crystallizing dish. Fill the outer beaker with iced-water to surround the vial and keep the dispersion cool during sonication.
5. Remove the lid from the vial and insert a 13 mm sonication probe into the vial, positioning it between the upper quarter and upper half.
6. Start sonication and run it for 16 min at 10% amplitude or 12 min at 20% amplitude, ensuring the sonication probe does not touch the vial walls.
7. After sonication, replace the lid on the vial.
8. Clean the sonication probe by sonicating it for 5 min in a 50:50 water-ethanol (96%) mixture. Afterward, rinse the probe in ethanol using a dispenser and a collection bottle. Allow the probe to air-dry in the fume hood.

Fig. 1 Setup for placement of sonication vial in ice

2.3.2 Storage Conditions of ENM

1. Use tightly sealed containers made of borosilicate glass or high-density polyethylene (HDPE). For photosensitive ENM, use amber-colored or opaque containers to protect from light.
2. Store in a dry environment, ideally inside a desiccator with silica gel or similar desiccant.
3. For oxidation-prone ENM (e.g., metallic NP), store under an inert atmosphere (e.g., nitrogen or argon).
4. Always follow manufacturer's specific storage recommendations.

2.3.3 Handling Recommendations of ENM

1. Minimize exposure to ambient air, particularly in the presence of humidity or reactive gases.
2. Handle ENM in a well-ventilated area or under a fume hood.
3. Avoid contamination between materials by using dedicated tools and containers.

2.4 Potentially Toxic Elements

2.4.1 PTE Stock Solution Preparation

1. Calculate the necessary mass (if working with a salt (*see* **Note 4**) to prepare your desired volume of stock solution. The molar mass of the salt can be used to calculate the exact amount of each compound required to prepare a specific concentration (*see* **Note 5**).

2. In a weighing boat, weigh the required amount of the PTE standard salt using an analytical balance (*see* **Note 6**).
3. Transfer the salt to a volumetric flask of the final desired volume using a powder funnel.
4. Add ultrapure water gradually to the flask, rinsing any remaining powder from the funnel into the flask with ultrapure water. Swirl gently until complete dissolution of the salt before bringing to final volume.
5. Once the salt is fully dissolved, bring the solution to the final volume with ultrapure water.
6. Mix thoroughly to ensure homogeneity. Label the flask with the PTE name, concentration, and date of preparation.
7. Always prepare fresh working solutions, prior to use, to avoid degradation or changes in concentration (*see* **Note 7**).

2.4.2 Storage and Handling Recommendations for PTE

1. PTE solutions should be stored in high-density polyethylene (HDPE) or borosilicate glass containers to avoid interactions with the container material.
2. Use dark-colored containers if the PTE are photosensitive (e.g., mercury).
3. Prepare and handle PTE solutions in a fume hood to avoid inhalation of vapors or aerosols.
4. Use dedicated pipettes and glassware to avoid contamination and ensure accurate measurements.

3 Methods

3.1 Batch Adsorption

There are several conditions that must be met for a robust ENM–PTE adsorption experiment to ensure accurate and reproducible results. Each independent experiment should be conducted in duplicate to account for variability, with a control containing only ultrapure water (Condition A) or cell culture media with or without added serum (Condition B), to establish a baseline for the elements under study. In parallel, single ENM (*see* **Note 8**) and single PTE incubations should be performed to assess potential adsorption to borosilicate glass, volatilization losses, and matrix interference, particularly in the case of complex matrices such as cell culture media (Fig. 2). It is recommended to perform three independent experiments to ensure statistical robustness.

1. Label borosilicate glass flasks according to each experimental condition.
2. Add 100 mL (*see* **Note 9**) of ultrapure water (Condition A) or cell culture media (Condition B) to each flask.

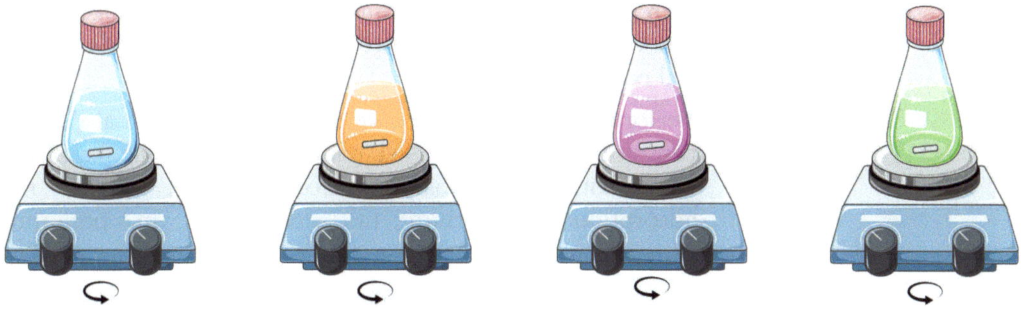

Fig. 2 Example of an experimental setup for an adsorption experiment

3. Insert a magnetic stir bar and place the flask on a magnetic stirrer (*see* **Note 10**). Begin stirring at ca. 100 rpm to ensure homogeneous mixing of ENM and PTE.
4. Monitor and record temperature and pH of the solution throughout the experiment.
5. Add the required volume of the ENM stock dispersion to reach the desired working concentration (*see* **Note 11**).
6. Add the required volume of the PTE stock solution to achieve the target concentration (*see* **Note 11**).
7. Start timing immediately after PTE addition.
8. Collect aliquots using a micropipette or a syringe connected to a silicone tube, keeping the stirrer on to maintain suspension homogeneity. Depending on the experimental design, follow the appropriate sampling strategy.

3.1.1 Kinetic Studies

1. When to use: Adsorption kinetics studies help determine how quickly interactions between ENM and PTE occur over time. Understanding the rate and extent of adsorption provides insight into the stability of the interaction and how rapidly contaminants might become bioavailable or exert toxic effects.
2. Immediately collect a 10 mL aliquot (t = 0 h) into a 15 mL centrifuge tube.
3. Subsequent samples should be withdrawn at predetermined time points (e.g., 5, 10, 30, 60, 90 min; 2, 4, 6, 12, 24, 48 h), depending on the expected adsorption rate (*see* **Note 11**).

3.1.2 Isotherm Studies

1. When to use: Adsorption isotherms help assess how the concentrations of PTE and ENM influence their interactions. Different ENM:PTE ratios can affect adsorption, which in

turn impacts the contaminants' solubility, stability, and bioavailability—key factors that shape the combined toxicological outcome.

2. Perform adsorption experiments using a fixed initial PTE concentration and vary ENM concentration, while keeping all other conditions constant. Allow sufficient time to reach adsorption equilibrium (as determined in preliminary experiments).
3. Collect duplicate 10 mL aliquots at equilibrium into 15 mL centrifuge tubes.
4. A minimum of 5–7 distinct equilibrium concentrations is recommended for reliable isotherm fitting.

3.1.3 Thermodynamic Studies

1. When to use: Thermodynamic studies are important because food products can be exposed to different temperatures during preparation. Addressing adsorption at 37 °C can be relevant to represent human body internal temperature.
2. Perform adsorption experiments at least three different temperatures (*see* **Note 13**). Monitor temperature during the complete duration of experiment (*see* **Note 14**).
3. Allow sufficient time to reach equilibrium at each temperature (as determined in preliminary experiments), and collect duplicate 10 mL aliquots once equilibrium is established.

3.1.4 pH Influence Studies

1. When to use: Adsorption studies at different pH levels help evaluate how pH influences the interaction between ENM and PTE. Changes in pH can alter surface charge, solubility, and binding affinity, all of which can affect adsorption behavior and ultimately impact toxicity outcomes.
2. Perform adsorption experiments, repeating the adsorption procedure at various pH values (*see* **Note 15**), adjusting the medium with 1 M HCl or 1 M NaOH as needed (*see* **Note 16**). Monitor and maintain the pH throughout the experiment, adjusting if necessary.
3. Allow adsorption to reach equilibrium (as determined in preliminary experiments) and collect duplicate 10 mL samples into 15 mL centrifuge tubes.

3.2 Separation: Filtration

1. Immediately after collection, filter the samples through a 0.22–0.45 μm filter, using a syringe and a syringe filter, and collect the filtrate to a new 15 mL labeled tube for subsequent PTE quantification and dissolved ENM, if desired (*see* **Note 17**).

2. If the sample matrix is ultrapure water (Condition A), acidify the sample to pH <2 with 15 μL HNO_3 65% and keep at 2–8 °C until analysis.

3.3 Digestion

If the sample matrix is cell culture media (Condition B), samples must be acid-digested prior to analysis.

3.3.1 Option 1: Microwave-Assisted Digestion

1. Place a 50 mL microwave digestion vessel in an analytical balance placed inside a hood. Transfer 9 mL of the cell culture media supernatant into the vessel (*see* **Note 18**). Record the exact mass.
2. Carefully add 1 mL of HNO_3 65% (*see* **Note 19**) to achieve an overall 10% HNO_3 concentration. Record the exact mass.
3. For quality control, include a blank digestion, replacing the cell culture media supernatant with ultrapure water or unused culture media) in each digestion batch to assess potential contamination from the digestion reagents. If available, digest a certified reference material (CRM) to verify the accuracy of the digestion protocol. Alternatively, spike the samples with a known concentration of the element(s) to be analyzed (*see* **Note 20**).
4. Close the vessels and place them in the rotor according to manufacturer instructions.
5. Transfer the vessels into the microwave digester and run the digestion program as following: (i) 15-min ramp to 175 °C; followed by (ii) 10-min hold at 175 °C; maximum power: 1000 W; pressure: ~800 psi.
6. Allow the vessels to cool down inside the microwave until they reach 80 °C.
7. Carefully remove the vessels and let them cool to room temperature in a fume hood (30–45 min).
8. After cooling, carefully remove the vessel lid and look at the cooled digested solutions. A transparent solution should be obtained (*see* **Notes 21** and **22**).
9. If dilution is required, either due to the composition of the cell culture or because PTE concentrations fall outside the calibration range, transfer the entire digested solution to a volumetric flask (e.g., 25 or 50 mL, adjust based on your analytical needs), rinsing the vessel with aliquots of ultrapure water, and then filling to the mark. Mix thoroughly by inverting the flasks several times. Otherwise, transfer the digested solution into clean sample vials suitable for analysis (e.g., ICP-MS, AAS, TXRF), and store them at the appropriate temperature (usually at 4 °C) until further analysis.

3.3.2 Option 2: Thermoblock or Water Bath Acid Digestion

1. Transfer 1–2 mL of cell culture media into an acid-washed, heat-resistant tube. Duplicate samples to evaluate the precision of the method.
2. For quality control, include a blank digestion in each digestion batch, replacing the cell culture media supernatant with ultrapure water or unused culture media, to assess potential contamination from the digestion reagents. If available, digest a certified reference material (CRM) to verify the accuracy of the digestion protocol. Alternatively, spike the samples with a known concentration of the element(s) to be analyzed (*see* **Note 20**).
3. Add the same volume of HNO_3 65% relative to the sample volume. For instance, for 1 mL of media, add 1 mL of HNO_3 65%.
4. Close the tubes and place them in a thermoblock or water bath set to 60 °C.
5. Allow the digestion to proceed for 18 h.
6. Optional: If visible residue remains after acid digestion, an additional oxidation step may be necessary. Following the initial heating, carefully remove the tubes and allow them to cool to room temperature under a fume hood. Then, add 100 μL of 30% hydrogen peroxide (H_2O_2) dropwise to each tube. Caution! The addition of H_2O_2 may result in vigorous bubbling and gas release. Allow the reaction to subside fully before loosely recapping the tubes—do not seal tightly to avoid pressure buildup, but ensure minimal volatilization. Return the tubes to the thermoblock or water bath at 60 °C and continue heating for 1 h, or until the solution becomes clear or only minimal residue remains and gas evolution has ceased.
7. After cooling, dilute the digested sample to 25–50 mL with ultrapure water (register the volume).
8. Transfer the digested solution into clean sample vials suitable for analysis (e.g., ICP-MS, AAS, TXRF), and store them at the appropriate temperature (2–8 °C) until further analysis.

4 Notes

1. While this protocol is designed to be applicable to a wide range of ENM and PTE, typical examples include TiO_2 or CeO_2 nanomaterials and metal ions such as Cd^{2+}, Hg^{2+}, or Pb^{2+}. Concentration ranges commonly used in adsorption studies vary depending on the system, but initial screening is often conducted using 1–25 mg/L for ENM are recommended to reduce the risk of ENM agglomeration and to maintain cell viability, if applicable. Special attention should be given to

ENM–protein interactions, which can affect dispersion and surface activity. For PTE, a range of 1–100 μg/L is typical for metals like Cd, Pb, or Hg to simulate low-dose chronic exposure. These values are meant as illustrative starting points and should be optimized based on the specific materials and justified by the goals of individual studies.

2. BSA solution can be filtered through a 0.22 μm filter to reduce the risk of bacterial contamination and extend its storage period.
3. Physicochemical Characterization of ENM: To support data interpretation and ensure reproducibility, it is strongly recommended to characterize the ENM using the following techniques: size and size distribution—Transmission Electron Microscopy (TEM), Dynamic Light Scattering (DLS); morphology—Scanning Electron Microscopy (SEM), Transmission Electron Microscopy (TEM); specific surface area—Brunauer-Emmett-Teller (BET) method; surface charge—Zeta potential measurements; crystallinity—X-ray Diffraction (XRD); dispersion stability—Polydispersity Index (PDI).
4. If starting from a solution, determine the required volume and add it directly to the volumetric flask. Bring the solution to the final volume with ultrapure water and mix thoroughly to ensure homogeneity.
5. Mass of compound (g) = (Molar mass of compound/Molar mass of ion) × Desired ion mass (g), where: (i) Molar mass of compound = molecular weight of the salt (e.g., $CdCl_2$); (ii) Molar mass of ion = atomic weight of the target ion (e.g., Cd^{2+}); (iii) Desired ion mass (g) = target concentration (g/L) × volume (L). If the compound is in hydrated form, the molar mass of the hydrate must be used.
6. Materials with a high degree of electrical insulation, such as glass and plastic, can become electrostatically charged. This can distort the weighing result of some salts. If experiencing this issue, we recommend using a metal container or a piece of aluminum foil.
7. If not used immediately, store solutions at 2–8 °C to prevent degradation or volatilization, especially for volatile PTE like mercury.
8. Although this procedure is primarily designed for adsorption of PTE onto ENM studies, it also serves as an ENM solubilization method [6], since a similar procedure can also be applied to study the release of ions from the ENM into solution. In co-exposure scenarios (ENM + PTE), both adsorption and solubilization phenomena may occur in parallel. It is important to note that ions released due to ENM solubilization are distinct from PTE adsorbed onto ENM and should be interpreted

separately. Following centrifugation, solubilized ENM-derived ions will remain in the supernatant and can be quantified to evaluate ENM solubility. When only ENM are present, the same workflow can be used exclusively for solubilization studies.

9. The initial assay volume must be planned according to the number and volume of aliquots to be withdrawn throughout the adsorption study. To ensure reliable and reproducible results, the removal of aliquots should not significantly alter the initial solid-to-liquid ratio or compromise the mixing efficiency. Ideally, the total volume withdrawn should not exceed 10–20% of the initial solution volume, particularly in kinetic studies. For example, if ten 5 mL aliquots are to be collected (totaling 50 mL), the starting volume should be at least 500 mL. This helps prevent particle settling and ensures consistent interaction conditions between PTE and ENM throughout the experiment.
10. In the case of magnetic ENM, magnetic bars cannot be used. In this case, an orbital shaker or an overhead stirrer should be used.
11. Determine the necessary volume of the stock solution to add by: Vstock = (Cfinal × Vfinal)/Cstock, where (i) Cstock = concentration of the stock solution; (ii) Cfinal = desired concentration of the working solution; (iii) Vfinal = final volume of the working solution.
12. Based on our experience, equilibrium is typically reached within 24 h or less [6]; however, the exact time required can vary considerably depending on the specific ENM, the target PTE, initial concentrations, pH, temperature, and other experimental parameters. It is recommended to perform a preliminary time-course study, using fewer time points over an extended period to accurately determine the equilibrium time for each specific ENM-PTE system.
13. For thermodynamic studies, we find it easier to maintain a stable temperature using a shaking incubator. However, since opening the incubator stops the shaking function, it is crucial to collect aliquots immediately to prevent ENM precipitation. Precipitation can lead to uneven ENM distribution, altering the sample composition and affecting experimental conditions. Depending on the number of samples being processed, it may be difficult to ensure immediate collection, so it is advisable to manually shake each flask before withdrawing an aliquot to maintain homogeneity.

14. Based on our experience, the set temperature of the incubator does not always match the actual temperature of the solutions. It is advisable to conduct a preliminary test to determine the incubator setting needed to achieve the desired solution temperature.
15. The choice of pH should reflect the experimental conditions you aim to replicate. pH 4–5 is useful for simulating acidic environments, such as those in the stomach or in scenarios involving acidic digestion or detoxification. pH 6–7 represents a neutral pH, common in many biological and environmental systems, including cell culture conditions and aquatic media.
16. Adjust pH using HCl or NaOH, and always check after adding other components, as they can shift pH. When testing adsorption across pH levels, pre-equilibrate your test solution to the desired pH before adding nanomaterials or metals to avoid unwanted reactions.
17. We recommend filtration as the preferred method for separating ENM or ENM–PTE complexes, as it is simple and quick. However, in some cases, filter clogging can occur, especially at higher ENM concentrations or in complex matrices such as cell culture media. Centrifugation can be used as an alternative method. To do so, centrifuge the sample at 30,000× *g* for 30 min immediately after collection. Then transfer the supernatant into another tube and centrifuge again at the same conditions, to separate the PTE and solubilized ENM from the ENM and ENM+PTE complexes. After centrifugation, carefully collect the supernatant using a micropipette or Pasteur pipette, ensuring that only the liquid phase is transferred. Place the supernatant into a labeled tube for subsequent PTE quantification (and dissolved ENM, if desired). Be cautious to avoid transferring any of the ENM or ENM-PTE complexes that may be in the pellet. Optional: The pellet can be used to determine the amount of adsorbed PTE or characterizing any residual ENM (by TEM, for example). It may also be valuable to analyze the pellet for potential phase changes or chemical interactions after adsorption.
18. Sample volume can be adjusted based on the analytical method and the expected concentration of the element(s). Accordingly, the volume of HNO_3 should also be adjusted. If high concentrations—outside the calibration curve—are expected, the sample can be diluted 5–10 times at this stage or after digestion, prior to analysis.
19. This is a general procedure, but we recommend that you ensure that the acid used in digestion does not interfere with the subsequent analysis of your desired element(s).

20. The mass of CRM or volume of spiking solution should be calculated based on the certified concentration of the target analyte(s). The resulting concentration in the sample matrix should approximate the expected analyte concentration in the sample and remain within the dynamic range of the calibration curve to ensure accurate quantification.
21. This digestion procedure is usually effective, but, as precaution, and to avoid clogging the ICP-MS nebulizer, a step can be added by filtering the solution through a microfilter with a pore size of 0.4 μm.
22. This procedure may require optimization depending on the composition of your cell culture media and the target elements. For complex samples, digestion conditions may need to be adjusted, potentially involving multiple steps or additional reagents. For example, TiO_2 nanomaterials require the addition of 50 μL of 40% HF after 1 mL of 65% HNO_3; however, HF should be avoided with Ce-containing samples, as it can cause precipitation.

Acknowledgments

This work was supported by FCT—Fundação para a Ciência e Tecnologia, I.P., through project MixIng-Tox with reference 2022.03491.PTDC and DOI identifier https://doi.org/10.54499/2022.03491.PTDC, and also projects with references UIDB/04750/2020 and LA/P/0064/2020 and respective DOI identifiers https://doi.org/10.54499/UIDB/04750/2020 and https://doi.org/10.54499/LA/P/0064/2020, and the projects with references UIDB/50011/2020, UIDP/50011/2020 and LA/P/0006/2020 and the respective DOI identifiers https://doi.org/10.54499/UIDB/50011/2020, https://doi.org/10.54499/UIDP/50011/2020 and https://doi.org/10.54499/LA/P/0006/2020. A.T.Reis thanks the financial support of FCT through Project MixIng-Tox with reference 2022.03491.PTDC and DOI identifier https://doi.org/10.54499/2022.03491.PTDC.

C.B. Lopes thanks the support of FCT—Fundação para a Ciência e Tecnologia, I.P. for researcher contract with DOI identifier https://doi.org/10.54499/2021.03739.CEECIND/CP1659/CT002.

References

1. Onyeaka H et al (2024) Preventing chemical contaminants in food: challenges and prospects for safe and sustainable food production. Food Control 155:110040
2. Bolan S et al (2024) The distribution, fate, and environmental impacts of food additive nanomaterials in soil and aquatic ecosystems. Sci Total Environ 916:170013

3. Thompson LA, Darwish WS (2019) Environmental chemical contaminants in food: review of a global problem. J Toxicol 2019(1): 2345283
4. Singh R et al (2023) Future of nanotechnology in food industry: challenges in processing, packaging, and food safety. Glob Chall 7(4): 2200209
5. Altemimi AB et al (2024) Application of nanoparticles in human nutrition: a review. Nutrients 16(5):636
6. Rosário F et al (2022) In vitro hepatotoxic and neurotoxic effects of titanium and cerium dioxide nanoparticles, arsenic and mercury co-exposure. Int J Mol Sci 23(5):2737
7. Rosário F et al (2020) Unravelling the potential cytotoxic effects of metal oxide nanoparticles and metal(loid) mixtures on A549 human cell line. Nanomaterials 10(3):447–467
8. Deng R et al (2017) Nanoparticle interactions with co-existing contaminants: joint toxicity, bioaccumulation and risk. Nanotoxicology 11(5):591–612
9. Naasz S, Altenburger R, Kühnel D (2018) Environmental mixtures of nanomaterials and chemicals: the Trojan-horse phenomenon and its relevance for ecotoxicity. Sci Total Environ 635:1170–1181
10. Forest V (2021) Combined effects of nanoparticles and other environmental contaminants on human health – an issue often overlooked. NanoImpact 23:100344
11. Zhang W et al (2020) Effect of metal oxide nanoparticles on the chemical speciation of heavy metals and micronutrient bioavailability in Paddy soil. Int J Environ Res Public Health 17(7):2482
12. Gatoo MA et al (2014) Physicochemical properties of nanomaterials: implication in associated toxic manifestations. Biomed Res Int 2014(1):498420
13. Lee YK et al (2015) Effect of the protein corona on nanoparticles for modulating cytotoxicity and immunotoxicity. Int J Nanomedicine 10:97–113
14. Tedja R et al (2012) Effects of serum adsorption on cellular uptake profile and consequent impact of titanium dioxide nanoparticles on human lung cell lines. ACS Nano 6(5): 4083–4093

Chapter 13

Human Ingestion of Micro-/Nanoplastics: A Simplified In Vitro Digestion Protocol for Cytotoxicity and Risk Assessment

Verónica Bastos and Helena Oliveira

Abstract

A comprehensive risk assessment of plastics' health impacts is imperative; however, significant gaps exist in toxicity models that closely simulate human ingestion. The biotransformation of micro−/nanoplastics (MNPLs) during the digestion process, along with their bioaccessibility, can significantly influence their toxicity. This chapter presents a simplified in vitro digestion protocol tailored for cytotoxicity assessment, providing a more realistic simulation of MNPLs ingestion. By enhancing the physiological relevance of laboratory models, this approach offers a valuable tool for improving the accuracy of human health risk evaluations related to plastic exposure.

Key words Digested micro- and nanoplastics, Ingestion pathway, In vitro digestion, Human exposure, Toxicological potential

1 Introduction

Global plastic (PL) production has escalated in recent decades, rising from just two million tons (Mt) in 1950 to 380 Mt. in 2015 [1]. By 2050, this figure is projected to reach an alarming 26,000 Mt. [1]. Plastics are now deeply embedded in our daily lives, from packaging (which accounts for 59% of global plastic production) to cosmetics, medicine, and countless other products [2]. However, the widespread use of plastics comes with significant environmental and health concerns, particularly related to microplastics (MPs) and nanoplastics (NPs), along with the chemicals they carry [3–6].

The presence of microplastics in the food chain has raised increasing concerns among international organizations about the potential adverse effects on human health [7]. Furthermore, recent studies have revealed alarming findings about the presence of MPs in various food and water sources [8]. For example, tap and bottled

Paula Alvito and Helena Oliveira (eds.), *Food Contaminants Toxicity*, Methods and Protocols in Food Science,
https://doi.org/10.1007/978-1-0716-5096-7_13,

water contain approximately 325 MPs per liter, while MPs have also been detected in fish, seafood, meat, salt, honey, sugar, fruits, vegetables, milk, and even beverages [9, 10]. It's estimated that humans ingest around 6000 MPs annually through these sources [8]. Indeed, MPs and NPs are now detected in human stools, providing evidence that ingestion is a significant pathway for human exposure to these tiny particles [11].

Despite the growing body of evidence, there is still a considerable gap in our understanding of the real effects of micro- and nanoplastics (MNPs) on human health. Most studies on the toxicity of microplastics focus on controlled experiments with specific types, sizes, and concentrations of particles [12–14]. These studies, however, often fail to reflect real-life conditions, including their biotransformation after ingestion, making it difficult to fully grasp the potential risks. This gap in knowledge has been acknowledged by scientific advisory bodies, including the Science Advice Policy of EU Academy [15], which has called for more comprehensive studies to assess the true impact of MNPs on human health and the environment.

To access the effects of MNPs under real gastrointestinal conditions, an in vitro digestion protocol should be implemented in realistic scenarios for human exposure aiding in the filling of toxicological knowledge gaps to enable a comprehensive risk assessment. Indeed, when simulating a realistic ingestion pathway, physicochemical modifications to MNPLs may occur, including the formation of a corona, which can impact their properties such as size, zeta potential, and ability to adsorb compounds [16], as well as their bioavailability and toxicological potential [17]. Additionally, gastric and intestinal fluids may enhance the solubilization of heavy metal content or other plastic additives [17].

Standardized digestion models, such as the INFOGEST protocol, are widely used to simulate food digestion under physiologically relevant conditions and have played a key role in harmonizing in vitro digestion studies across laboratories [17, 18]. However, INFOGEST was primarily developed for food matrices and includes multiple complex steps, making it less suitable for high-throughput toxicological assessments involving environmental samples.

Recent studies have demonstrated that gastrointestinal digestion can markedly alter the physicochemical properties and biological behavior of nanoplastics, affecting their cellular uptake and toxicity profiles [19, 20]. Despite this, the complexity of existing digestion protocols often limits their routine use in toxicology-focused research.

To address this gap, we introduce a simplified, bottom-up digestion protocol specifically designed for cytotoxicity assessment (Fig. 1). This method simulates exposure to gastric acid followed by neutralization with intestinal fluid, both containing relevant

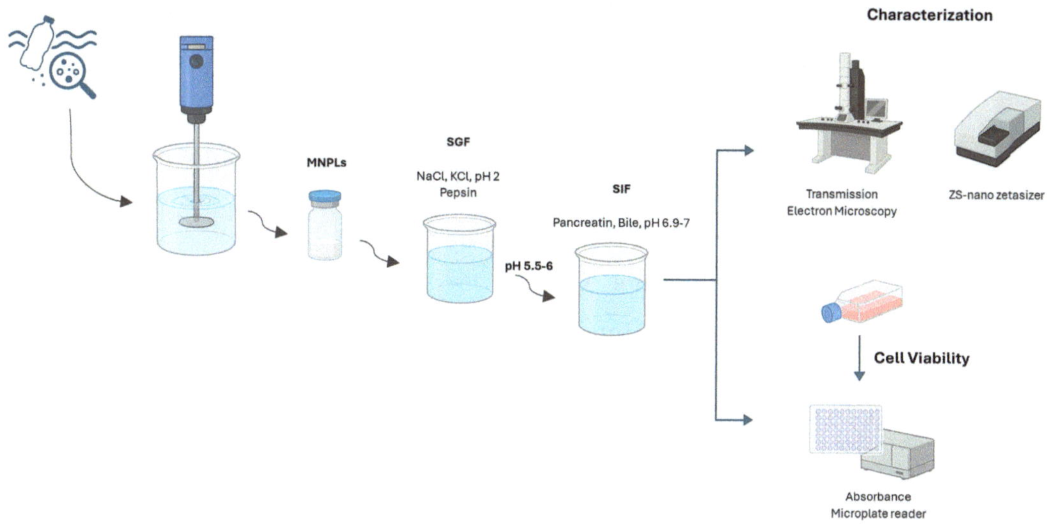

Fig. 1 Schematic representation of the simplified *in vitro* digestion protocol to assess MNPs cytotoxicity

digestive enzymes. Unlike traditional protocols, our approach begins with real-world plastic items and follows a streamlined process from digestion to cytotoxicity testing of micro- and nanoplastics. This accessible, time-efficient methodology aims to enhance the physiological relevance of toxicological models, ultimately supporting more realistic health risk assessments and informing regulatory and public health decision-making.

2 Materials

Sodium chloride (NaCl), potassium chloride (KCl), sodium bicarbonate ($NaHCO_3$), hydrochloric acid (HCl, 1 mM), bile extract (e.g., porcine or bovine), pepsin (from porcine gastric mucosa), pancreatin (from porcine pancreas), ultra-pure or Milli-Q water. Filtration for solution sterilization: 0.22 μm sterile syringe filters (Millipore®); filtration of plastic particles used at Subheading 3.1 of the methodology: filters with 2105 grade (particle retention 12–15 μm).

2.1 Preparation of Solutions

Prepare all solutions using analytical grade reagents. Prepare and store all reagents at room temperature (unless indicated otherwise).

2.1.1 Preparation of Digestive Fluids (Adapted from [19, 21])

Simulated Gastric Fluid (SGF).

1. For a final volume of 10 mL:
   - 140 mM NaCl.
   - 5 mM of KCl.
2. Adjust pH to 2 with sterile 0.10 M HCl.

3. Sterilize the solution by autoclaving (120 °C, 20 min).
4. Store at RT.

Porcine pepsin solution:

5. Concentration: 25 mg/mL (800–2500 U/mg protein).
6. Solvent: 0.1 M HCl.
7. Sterilize the solution by filtration.
8. Store at 4 °C.

Simulated Intestinal Fluid (SIF).

1.40 mg/mL pancreatin.

8 mg/mL bile.

9. Dissolve in MiliQ water (previously sterilized by autoclaving).
10. Sterilize the solution by filtration.
11. Store at 4 °C.

3 Methods

Top-down method to produce MNPs by using an Ultra-Turrax disperser [22] (*see* **Note 1**):

3.1 Plastics Fragmentation

1. Plastic materials from different daily plastic items with different forms and shapes are cut into 2 cm pieces.
2. Place 6 g of plastic materials into a high-walled 400 mL beaker.
3. Add 115 mL of acetone to the beaker in a fume hood.
4. Cool the beaker using a water and ice bath to minimize solvent evaporation and control temperature during dispersion.
5. Disperse the polymer using a T18 digital Ultra-Turrax® disperser at 18,000 rpm for 10 min.
6. Immediately after dispersion, clean the disperser tool using 5 mL of fresh acetone to remove any adhered plastic debris.
7. Filter the resulting dispersion to remove larger plastic debris.

   Solvent Exchange (Acetone to Water) in a fume hood:

8. Evaporate the acetone until the dispersion volume is reduced to approximately 30 mL.
9. Add 115 mL of purified water to the concentrated dispersion.-Note: The liquid will turn turbid immediately due to the water-acetone exchange and polymer aggregation.
10. Continue evaporating to remove any remaining acetone.

    Final Filtration:

11. Filter the aqueous dispersion again to remove aggregates formed during the evaporation process.

MNPs are ready to be characterized (go the Subheading 3.3) named as non-digested MNPs.

12. The aqueous suspension of MNPs must be freeze-dried prior to the digestion process.

3.2 Simplified In Vitro Digestion (Adapted from [19, 21])

Prepare a blank solution that undergoes all the digestion steps except it does not contain the MNPs. This solution should then be used to assess cytotoxicity in Subheading 3.4.

3.2.1 Simulation of Gastric Digestion (See Note 2)

1. Disperse 0.120 g of MNPs into 3 mL of the simulated gastric fluid (SGF) solution.
2. Add 0.150 mL of sterile porcine pepsin solution.
3. Confirm pH and adjust to 2 with sterile 0.10 M HCl if needed.
4. Incubate samples: On rocking platform (Model RP-50, Rockville, MD, USA); 55 oscillations/min; 1 h at 37 °C.

3.2.2 Transition to Intestinal Digestion

1. Adjust pH to 5.5–6.0 using sterile 0.10 M $NaHCO_3$ (previously sterilized by filtration).

3.2.3 Simulation of Intestinal Digestion

1. Add 0.150 mL of the simulated intestinal fluid (SIF) solution.
2. Adjust pH to 6.9–7 using sterile 0.10 M $NaHCO_3$.
3. Incubate samples: On rocking platform (Model RP-50, Rockville, MD, USA); 55 oscillations/min; 1 h at 37 °C.
4. Bring volume to 4.5 mL using the sterile solution of 140 mM NaCl and 5 mM KCl at pH 6.7.
5. Label the obtained samples as digested plastics and store at 4 ° C.

3.3 Characterization of Digested and Non-digested Plastics

1. Evaluate size distribution and shape of digested plastics through TEM/SEM images and their hydrodynamic size was determined by DLS (*see* **Note 3**).

3.3.1 Transmission Electron Microscopy (TEM) [19]

1. Dilute particle dispersions in distilled water to a 200 μg/mL concentration.
2. Mount the diluted dispersions on carbon-coated TEM grids.
3. Visualize particles using the JEOL JEM-1400 instrument (JEOL LTD, Tokyo, Japan).
4. To determine mean size and distribution: Measure 100 randomly selected particles, use ImageJ software for analysis.

3.3.2 Scanning Electron Microscopy (SEM) [19]

1. Dilute particle dispersions in distilled water to a final concentration of 200 μg/mL.
2. Place a drop of each dispersion onto a silica grid.
3. Allow the samples to evaporate overnight.
4. Visualize NPL characteristics using SEM Zeiss Merlin device (Zeiss, Oberkochen, Germany).
5. Record the SEM signals.
6. Analyze recorded signals using INCA Energy software.

3.3.3 Dynamic Light Scattering (DLS) [23]

Characterize the obtained digested products considering the hydrodynamic sizes of the MNPs before and after in vitro gastrointestinal digestion by DLS using a ZS-nano zetasizer (Malvern Panalytical, Malvern, Great Britain):

1. Sonicate suspensions for 1 min using a sonication bath.
2. Load the sonicated suspensions into 1.5 mL polystyrene cuvettes.
3. For each sample, obtain at least 5 autocorrelation curves.
4. Measure in triplicate.
5. Include a medium blank with each measurement set to assess background particulate matter.

3.4 Cytotoxicity Assessment (See Note 4)

Evaluate cytotoxicity by exposing digested products to in vitro cell cultures:

1. Briefly, maintain cell lines in cell culture medium with respective additives suitable for each cell line (usually supplemented with 10% (v/v) fetal bovine serum (FBS), 2 mM L-glutamine, 1% pen/strep (100 U/mL penicillin, 100 μg/mL streptomycin), and 2.5 μg/mL fungizone at 37 °C, under a humidified atmosphere containing 5% CO_2).
2. Seed the cells in 96-well plates, allow to adhere for 24 h (if working with adherent cell lines), and expose to different concentrations of digested and non-digested plastics (concentrations up to 1 mg/mL may be considered, based on daily intake estimations [24–26]) during different periods (24, 48 h).
3. Include the following controls:
   - A negative control consisting of cells exposed to culture media only.
   - A blank control, consisting of digestion fluids without MNPs, which underwent the same digestion protocol and was applied at the same volume as the highest MNP concentration tested.

Assess cells viability through WST-8 assay [27] (*see* **Note 5**):

4. Add 10 μL of WST-8 solution to each well.
5. Incubate the plate for 2 h at 37 °C.
6. Before reading the plate, it is important to mix gently on an orbital shaker for 1 min to ensure homogeneous distribution of color.
7. Measure the absorbance at 460 nm using a microplate reader.
8. Calculate Cell Viability with the equation:

$$\% \text{ Cell viability} = (\text{Atreatment} - \text{Ablank})/(\text{Acontrol} - \text{Ablank}) \times 100\% \quad (\text{where, } \text{A} = \text{absorbance}).$$

4 Notes

1. Although the same fragmentation protocol was applied to all plastic types, it is important to note that polymers differ in their physical and chemical properties—such as brittleness, crystallinity, and solvent resistance—which can influence fragmentation efficiency and particle characteristics. For instance, brittle materials like PS fragment more readily under shear, while ductile polymers like HDPE may yield larger or more irregular particles. These differences are acknowledged as potential sources of variation and should be considered in the interpretation of toxicological outcomes.
2. Simulated salivary fluid (SSF) with salivary amylase could be used to digest MNPLs in starch-containing food and to mimic the oral phase by using simulated salivary fluid (SSF) [17].
3. Different techniques could be used to characterize digested plastics: MNPLs composition could be characterized with Fourier transform infrared spectroscopy (FITR) and Raman spectroscopy [28]. Protein corona composition could be quantified using LC–MS–MS-based proteomics as reported by [23]. Also, BCA (bicinchoninic acid) protein assay kit could be used to assess the adsorbed protein contents on surface of MNPLs [16].
4. Intracellular localization of digested plastics could be evaluated through flow cytometer and confocal microscopy by using for, e.g., fluorescent yellow counterpart as shown by Vela et al. [19].
5. Cytotoxicity could be assessed with various cell viability assays as for, e.g., MTT, LDH, Trypan Blue, or related. Various endpoints could be used to evaluate cytotoxicity beyond cells viability assays, such as oxidative stress through ROS DCFDA assay, blockage of cytokinesis through CB Micronucleus assay, apoptosis though Annexin V, among others [29].

Acknowledgments

The authors acknowledge the financial support to UID Centro de Estudos do Ambiente e Mar (CESAM) by national funds through FCT—Fundação para a Ciência e a Tecnologia I.P.—under the project/grant UID/50017/2025 (https://doi.org/10.54499/UID/50017/2025) and LA/P/0094/2020 (https://doi.org/10.54499/LA/P/0094/2020). This work was supported by the WIN-PLASTIC project—"World in Plastic—Impact of plastic and bioplastic ingestion on the gut and human health project" (COMPETE2030-FEDER-00843100) and funded by Innovation and Digital Transition Program (COMPETE2030) and the Regional Program for Lisbon (LISBOA 2030) through their FEDER component. The Portuguese Foundation for Science and Technology (FCT) is also acknowledged for the research contract of V. Bastos (https://doi.org/10.54499/2022.05740.CEECIND/CP1720/CT0031) and H. Oliveira (https://doi.org/10.54499/CEECIND/04050/2017/CP1459/CT0023).

References

1. Geyer R, Jambeck JR, Law KL (2017) Production, use, and fate of all plastics ever made. Sci Adv 3. https://doi.org/10.1126/SCIADV.1700782
2. Pazienza P, De Lucia C (2020) The EU policy for a plastic economy: reflections on a sectoral implementation strategy. Bus Strateg Environ 29:779–788. https://doi.org/10.1002/bse.2445
3. Verla AW, Enyoh CE, Verla EN, Nwarnorh KO (2019) Microplastic-toxic chemical interaction: a review study on quantified levels, mechanism and implication. SN Appl Sci 1. https://doi.org/10.1007/s42452-019-1352-0
4. Meng J, Xu B, Liu F, Li W, Sy N, Zhou X et al (2021) Effects of chemical and natural ageing on the release of potentially toxic metal additives in commercial PVC microplastics. Chemosphere 283. https://doi.org/10.1016/J.CHEMOSPHERE.2021.131274
5. Prata JC, Reis V, Paço A, Martins P, Cruz A, da Costa JP et al (2020) Effects of spatial and seasonal factors on the characteristics and carbonyl index of (micro)plastics in a sandy beach in Aveiro, Portugal. Sci Total Environ 709. https://doi.org/10.1016/J.SCITOTENV.2019.135892
6. Rillig MC, Kim SW, Kim TY, Waldman WR (2021) The global plastic toxicity debt. Environ Sci Technol 55:2717–2719. https://doi.org/10.1021/ACS.EST.0C07781
7. FAO (2023) The impact of microplastics on the gut microbiome and health – a food safety perspective, Food Safety and Quality Series, vol 21, 21st edn. FAO, Rome. https://doi.org/10.4060/cc5294en
8. Kosuth M, Mason SA, Wattenberg EV (2018) Anthropogenic contamination of tap water, beer, and sea salt. PLoS One 13:e0194970. https://doi.org/10.1371/journal.pone.0194970
9. Food Safety Authority E (2016) Presence of microplastics and nanoplastics in food, with particular focus on seafood EFSA Panel on Contaminants in the Food Chain (CONTAM). https://doi.org/10.2903/j.efsa.2016.4501
10. Hirt N, Body-Malapel M (2020) Immunotoxicity and intestinal effects of nano- and microplastics: a review of the literature. Part Fibre Toxicol 17:1–22. https://doi.org/10.1186/S12989-020-00387-7
11. Schwabl P, Koppel S, Konigshofer P, Bucsics T, Trauner M, Reiberger T et al (2019) Detection of various microplastics in human stool. Ann Intern Med 171:453–457. https://doi.org/10.7326/M19-0618
12. Khan A, Jia Z (2023) Recent insights into uptake, toxicity, and molecular targets of microplastics and nanoplastics relevant to human health impacts. IScience 26:106061. https://doi.org/10.1016/J.ISCI.2023.106061

13. Niu H, Liu S, Jiang Y, Hu Y, Li Y, He L et al (2023) Are microplastics toxic? A review from eco-toxicity to effects on the gut microbiota. Meta 13:739. https://doi.org/10.3390/METABO13060739/S1
14. Ali N, Katsouli J, Marczylo EL, Gant TW, Wright S, Bernardino de la Serna J (2024) The potential impacts of micro-and-nano plastics on various organ systems in humans. EBioMedicine 99:104901. https://doi.org/10.1016/J.EBIOM.2023.104901
15. European Commission D-G for R and I (2019) A scientific perspective on microplastics in nature and society – Scientific Advice Mechanism. https://doi.org/10.26356/microplastics
16. Liu S, Wu X, Gu W, Yu J, Wu B (2020) Influence of the digestive process on intestinal toxicity of polystyrene microplastics as determined by in vitro Caco-2 models. Chemosphere 256:127204. https://doi.org/10.1016/J.CHEMOSPHERE.2020.127204
17. Brodkorb A, Egger L, Alminger M, Alvito P, Assunção R, Ballance S et al (2019) INFOGEST static in vitro simulation of gastrointestinal food digestion. Nat Protoc 14:991–1014. https://doi.org/10.1038/s41596-018-0119-1
18. Minekus M, Alminger M, Alvito P, Ballance S, Bohn T, Bourlieu C et al (2014) A standardised static in vitro digestion method suitable for food – an international consensus. Food Funct 5:1113–1124. https://doi.org/10.1039/C3FO60702J
19. Vela L, Villacorta A, Venus T, Estrela-Lopis I, Pastor S, García-Rodriguez A et al (2023) The potential effects of in vitro digestion on the physicochemical and biological characteristics of polystyrene nanoplastics. Environ Pollut 329:121656. https://doi.org/10.1016/J.ENVPOL.2023.121656
20. Paul MB, Böhmert L, Thünemann AF, Loeschner K, Givelet L, Fahrenson C et al (2024) Influence of artificial digestion on characteristics and intestinal cellular effects of micro-, submicro- and nanoplastics. Food Chem Toxicol 184:114423. https://doi.org/10.1016/J.FCT.2023.114423
21. Mahler GJ, Shuler ML, Glahn RP (2009) Characterization of Caco-2 and HT29-MTX cocultures in an in vitro digestion/cell culture model used to predict iron bioavailability. J Nutr Biochem 20:494–502. https://doi.org/10.1016/J.JNUTBIO.2008.05.006
22. Hildebrandt J (2023) Studies on nanoplastics for the preparation of reference materials. Dissertation. University of Potsdam and Bundesanstalt für Materialforschung und -prüfung (BAM). https://doi.org/10.25932/publishup-61710
23. Brouwer H, Porbahaie M, Boeren S, Busch M, Bouwmeester H (2024) The in vitro gastrointestinal digestion-associated protein corona of polystyrene nano- and microplastics increases their uptake by human THP-1-derived macrophages. Part Fibre Toxicol 21:1–26. https://doi.org/10.1186/S12989-024-00563-Z
24. Mason SA, Welch VG, Neratko J (2018) Synthetic polymer contamination in bottled water. Front Chem 6:389699. https://doi.org/10.3389/FCHEM.2018.00407/BIBTEX
25. Cox KD, Covernton GA, Davies HL, Dower JF, Juanes F, Dudas SE (2019) Human consumption of microplastics. Environ Sci Technol 53:7068–7074. https://doi.org/10.1021/ACS.EST.9B01517/ASSET/IMAGES/LARGE/ES-2019-015177_0004.JPEG
26. Qian N, Gao X, Lang X, Deng H, Bratu TM, Chen Q et al (2024) Rapid single-particle chemical imaging of nanoplastics by SRS microscopy. Proc Natl Acad Sci USA 121:e2300582121. https://doi.org/10.1073/PNAS.2300582121/SUPPL_FILE/PNAS.2300582121.SAPP.PDF
27. Cayman Chemical Company. WST-8 Cell Proliferation Assay Kit 2016. www.caymanchem.com. Accessed 6 May 2025
28. Da Costa JP, Nunes AR, Santos PSM, Girão A V, Duarte AC, Rocha-Santos T, et al. Degradation of polyethylene microplastics in seawater: insights into the environmental degradation of polymers 2018;53:866–875. https://doi.org/10.1080/10934529.2018.1455381, J Environ Sci Health A
29. Bastos V, Ferreira de Oliveira JMP, Brown D, Jonhston H, Malheiro E, Daniel-da-Silva AL et al (2016) The influence of citrate or PEG coating on silver nanoparticle toxicity to a human keratinocyte cell line. Toxicol Lett:249. https://doi.org/10.1016/j.toxlet.2016.03.005

Chapter 14

Porcine Explants and Slices: Original Models to Analyze the Effects of Mycotoxins

Philippe Pinton, Delphine Payros, Laura Soler-Vasco, and Isabelle P. Oswald

Abstract

In addition to in vitro approaches that use intestinal or liver cells in culture, ex vivo approaches represent a valuable model for evaluating the gastrointestinal toxicity of food contaminants. This chapter provides a comprehensive description of the preparation of punch-excised intestinal explants and precision-cut intestinal or liver slices. It illustrates the application of these techniques to the toxicity of mycotoxins, natural contaminants that are very prevalent in food and feed. These toxins are produced by toxigenic filamentous fungi, predominantly belonging to the genera *Aspergillus*, *Fusarium*, and *Penicillium*.

Key words Cultured cells, Ex vivo models, Intestine, Liver, Explants, Slices, Food/feed contaminants, Mycotoxins

1 Introduction

The gastrointestinal tract is the primary target for food and feed contaminants such as mycotoxins and has been the subject of many studies, mainly conducted using cell culture tests or in vivo assays. The culture of pieces of tissue or organ, obtained from an organism and whose cells are kept alive [1], represents an alternative model intermediate between in vitro and in vivo approaches [2, 3]. This method includes the preparation of punch-excised explants or precision-cut slices. These mini models of organs preserve the three-dimensional structure of interest, retain cell-cell and cell-extracellular matrix interactions, and preserve the resident cells present, including immune cells. They are thin enough to allow sufficient oxygen and nutrient supply to the inner cell layers [4]. Providing from diverse animal species, they offer the advantages of obtaining hundreds of samples from the same biological source, enabling toxic compound screening studies and reducing study costs. Data analysis may also benefit from the fact that

Paula Alvito and Helena Oliveira (eds.), *Food Contaminants Toxicity*, Methods and Protocols in Food Science,
https://doi.org/10.1007/978-1-0716-5096-7_14,

isogenic controls are present for each treatment, enabling pairwise analysis. Ultimately, these models adhere to the 3Rs principle (replacement, reduction, and refinement), which encourages alternatives to reduce the use of animals for scientific purposes.

It should be noted that these models are not without limitations. The lifespan of the samples is limited to a few hours for intestinal samples and a few days for liver samples. In the case of intestinal samples, the apical and basolateral sides are not distinguished, and the microbiota and blood circulation are not considered.

Initially, the precision cutting technique was applied to solid organs such as liver and kidney, but not to the intestine. As a hollow, tender, long, highly differentiated, and delicate organ, it was more difficult to cut [5]. For this reason, intestinal samples were first produced with a coring tool and cut directly into the intestinal wall, thus obtaining the full thickness of the intestinal wall, as described below for pig [6].

The use of a tissue slicer instrument was made possible by embedding intestinal segments in low-melting agarose, allowing improved tissue reproducibility and viability, as briefly described here [4]. Precision-cut intestinal slices (PCIS) were first developed in rats [7], then applied to mice and chickens. Punch-excised explants and precision-cut tissue slices (PCTS) permit the acquisition of multiple data points, including those derived from the explant/slices themselves and from the culture medium [8]. They have been widely explored, not only in the field of cell metabolism in healthy and cancer cells, but also in drug transport, metabolism, and toxicity of drugs, fibrosis, and viral infections [5]. The main advantages and disadvantages of punch-excised explants and precision-cut tissue slices, compared with in vitro and in vivo models, are summarized in Table 1.

Mycotoxins are natural food and feed contaminants produced by filamentous toxigenic fungi, mainly from *Aspergillus*, *Fusarium*, and *Penicillium* genus. Based on their occurrence and/or toxicity, regulations have been established in Europe for aflatoxin B1 (AFB1), deoxynivalenol (DON), fumonisin B1 (FB1), ochratoxin (OTA), patulin (PAT), zearalenone (ZEN), and ergot alkaloids (EAs). Moreover, future research must address the study of less documented "modified forms" and "emerging" mycotoxins [9]. Due to their different chemical structures, mycotoxins have different modes of action and exert various deleterious effects on human or animal health, in particular on the gastrointestinal tract [10].

This chapter outlines the main steps to prepare and cultivate punch-excised explants and precision-cut slices and delineates the application of these models to assess the intestinal and hepatic toxicity of mycotoxins, including the interactions between multiple mycotoxins that can be present in food or feed [11].

Table 1
Main advantages and disadvantages of punch-excised explants and precision-cut tissue slices compared with in vitro and in vivo models

	Advantages	Disadvantages
Ex vivo explants and slices	Multiplicity of explants/slices from the same animal Reduced variability (explant from the same animals being control and treated) Explants/slices from various organs from the same animal Maintenance of the different cell types in their normal environment Preservation of intercellular and cell-matrix interactions Preservation of morphological and architectural structures Multiple investigations (endogenous metabolism, biotransformation, transport of xenobiotics, toxicology)	Animals still required Lifespan limited, for some organs No zootechnical performances Absence of vascular system and systemic immune system For intestine, few studies in presence of microbiota
In vitro cell culture	Single cell type Limited availability (transformed cell lines) Model of choice for mechanistic studies (multiple reagents available, various genome modification tools)	Animals still required (primary cells) Finite life span (primary cells) Possible dedifferentiation with multiple passages Cancer cell lines frequently display DNA mutations
In vivo experiments	Inter-organ communication preserved Reliable approach for observing specific biological effects in a target species or in a model species for human Zootechnical performances Complete microbiota Complete cell-cell and cell-matrix interactions	Specific facilities and authorizations required Ethical issues Interindividual variability Complexity of the response Limited repetition number Costly

In the future, organs other than the gastrointestinal tract could be considered, such as urinary tract, kidney and bladder (for instance, to study the toxicity of ochratoxin A, a mycotoxin that accumulates in this organ), or lung (to study the effects of airborne mycotoxins). Future improvements should focus on how these models can better reproduce chronic exposure.

2 Materials

2.1 Animals

Most of the studies presented in this chapter were performed using samples derived from pigs. Indeed, the pig serves as both as a target species for mycotoxins due to the potential contamination of its diet and as a model species, relevant for human toxicology [12].

2.2 Reagents

1. Complete Krebs-Henseleit Buffer (KHB): supplemented with $NaHCO_3$ (2.1 g/L) and $CaCl_2$, $2H_2O$ (0.373 g/L), supplemented with penicillin (100 UI/ mL), streptomycin (100 μg/ mL), always maintained in ice and oxygenated (95% O_2, 5% CO_2).
2. Complete Williams medium E (WME): calf serum (1%), gentamycin (2 mM), penicillin (100 UI/mL), streptomycin (100 μg/mL), always maintained in ice and oxygenated (95% O_2, 5% CO_2).
3. Paraffin-filled petri dishes.
4. Low-melting agarose (3%) dissolved in WME.
5. Paraformaldehyde (PFA) (4%) in sodium phosphate buffer.

2.3 Equipment

Krumdieck tissue slicer, tissue coring tool, tissue embedding unit (Alabama, Munford, AL, USA).

3 Methods

3.1 Punch-Excised Intestinal Explants

1. From euthanized pigs, collect intestinal tissue samples (8 cm long) from different anatomical regions of interest (*see* **Note 1**).
2. Store them as quick as possible and for less than 1 h in WME.
3. Flush and rinse with WME to remove luminal debris and store in WME.
4. Cut open the intestinal samples in a longitudinal direction and hold stretched with needles on a paraffin-filled petri dish.
5. Punch out explants using a biopsy punch (diameter of 6–10 mm) and store them in WME.
6. Incubate in 6-well plates four explants per well on sterile foam biopsy pads (sponge) under gentle agitation at 80–90 rpm (37 °C, 5% CO_2) in 2.5 mL WME for 30 min to eliminate the debris. Alternatively, intestinal explants can be mounted in Ussing chamber to assess intestinal barrier function (*see* **Note 6**).
7. Then transfer the sponges and explants into plates containing WME supplemented with mycotoxins of interest (*see* **Note 4**).
8. After incubation (up to 8 h), collect and snap-freeze in liquid nitrogen to assess gene expression (*see* **Note 7**) and protein or metabolite abundance (*see* **Notes 8** and **9**), or fix in PFA for 24 h and then store in ethanol 70% until paraffin embedding prior to histomorphology evaluation (*see* **Note 12**). Collect, centrifuge, and freeze cell culture medium (−80 °C) for analysis of secretome (*see* **Note 10**).

3.2 Precision-Cut Intestinal Slices

1. Unlike rodent precision-cut intestinal slices, the preparation of pig ones is different due to the thickness of muscle layer and circumference of the lumen—too large to cut an entire ring of intestine.
2. Flush intestinal samples through the lumen with KHB, carefully remove adhering fatty tissue and muscle layer, and cut the segments of interest into 3-cm parts.
3. Dissect 0.5 × 0.2 cm strips, embed in low-melting agarose (3%) and, once solidified, cut into 300 μm thick slices before their transfer into KHB.
4. For smaller animals such as rats close one end of each segment with surgical thread and fill it with 3% agarose solution at 37 °C until the intestine reaches its normal diameter. Close opposite end and immediately transfer the filled segment to KHB to allow the agarose to solidify.
5. When the agarose has solidified, cut the intestine segment into two halves using a scalpel blade. Place a pin in the intestinal lumen avoiding touching the tissue and transfer the segment to the pre-cooled mold of the embedding unit, with the pin upward. Holding it, fill the mold with liquid agarose at 37 ° C, avoiding the inclusion of air bubbles. Allow the agarose to solidify for 1 min, then remove the pin.
6. Place the agarose cylinder containing the tissue into the slicer device such as Alabama slicer [13]. Cut slices not less than 100 μm thick, otherwise damaged cells on the cut edges will outnumber healthy cells (*see* **Note 3**).
7. Regularly collect the slices in a beaker containing KBH. Select slices according to their appearance and transfer them into the chosen oxygenated culture medium at 37 °C. Renew the slicer buffer every 20 min, and change the knife when the quality decreases.
8. Transfer selected, good quality slices round, equally thick at all sides with smooth edges into 12-well plates (one slice per well) in 1.3 mL WME containing the mycotoxins of interest and incubate (37 °C, 5% CO_2) (*see* **Note 4**).
9. A pre-incubation/regeneration stage of 1 h is recommended (*see* **Note 5**).
10. After incubation (suggested up to 72 h), snap-freeze in liquid nitrogen to assess gene expression (*see* **Note 8**) and protein or metabolite abundance (*see* **Notes** 7 and **9**), or fix in PFA for 24 h and then store in ethanol 70% until paraffin embedding prior to histomorphology evaluation (*see* **Note 12**). Collect, centrifuge, and freeze cell culture medium (−80 °C) for analysis of secretome (*see* **Note 10**).

3.3 Precision-Cut Liver Slices

1. From euthanized animals, collect a lobe of liver (*see* **Note 2**), flush it with KHB or University of Wisconsin (UW) solution until color changes from dark brown to permanent light brown and store it in KHB (*see* **Notes 1** and **2**).
2. Prepare 6–10 mm diameter cores with hand-driven biopsy punches or electrical drills and store in KHB. Cut 250 μm thick slices with a slicer device (*see* **Note 3**).
3. Transfer selected, good quality slices round, equally thick at all sides with smooth edges into 12-well plates (one slice per well) in 1.3 mL WME containing the mycotoxins of interest and incubate (37 °C, 5% CO_2) (*see* **Note 4**).
4. A pre-incubation/regeneration stage of minimum 1 h is recommended prior to exposure to mycotoxins (*see* **Note 5**).
5. After incubation (suggested up to 72 h), snap-freeze in liquid nitrogen to assess gene expression (*see* **Note** 7), protein or metabolite abundance (*see* **Notes 8** and **9**), and mycotoxins biotransformation (*see* **Note 11**), or fix in PFA for 24 h and then store in ethanol 70% until paraffin embedding prior to histomorphology evaluation (*see* **Note 12**). Collect, centrifuge, and freeze cell culture medium (−80 °C) for analysis of secretome (*see* **Note 10**).

4 Notes

1. Sampling conditions. If more than one organ is harvested from an animal, first excise the intestine, as it is sensitive to warm ischemia. Then store it in KHB. The liver, after recovery, should be perfused and stored in a cold organ preservation solution such as University of Wisconsin (UW) solution until preparation of slices. The perfusion of rat or mouse livers does not appear necessary as it can deteriorate the viability during incubation [14].
2. Liver slices. Liver samples can be obtained from different animal species or potentially following human surgery. Precision-cut liver slices (PCLS) are an alternative model to isolated hepatocyte culture and include the different fractions of liver (hepatocytes, Kupffer, hepatic stellate, fibroblasts, endothelial and biliary epithelial cells).
3. Optimized coring and tissue sections. Prepare cores using a rotating tissue coring tool. Cores should be cylindrical, with similar diameters at the two ends. To avoid damages, manipulate the cores using a spatula instead of forceps. The thickness commonly used for PCLS is 250 μm. It allows access to nutrients and oxygen for the inner cell layers. This is not the case

with slices thicker than 300 μm. If required, the thickness can be reduced to 100 μm but it would result in low viability rates due to too high a ratio of damaged to undamaged cells [4]. Select slices showing a good aspect (equal thickness and weight, uniform color, and smooth edges) and transfer into KHB (intestinal slices) or fresh ice-cold UW (liver slices) using a spatula [4].

4. Composition of storage and culture media, conditions of incubation. Intestinal samples are preserved in KHB, while UW solution is a recommended alternative for liver. The widely used culture conditions for intestinal explants or slices and liver slices are with WME and incubation with 5% CO_2 and high concentration of O_2. Indeed, a high O_2 concentration (80%) favors slices viability [8] and allows an upregulation of a large number of genes related to liver physiology [15]. In addition, to improve viability, it is recommended to incubate plates containing slices under gentle agitation (80–90 rpm).
5. Pre-incubation regeneration stage. During preparation, slices are subjected to mechanical and oxidative stress; a pre-incubation period of 1 h improves tissue viability [16]. After collection, slices are rapidly washed in cold WME before transfer (one per well) to a pre-warmed culture plate. After 1 h, they are incubated in WME containing the mycotoxins of interest. After treatment, intestinal or liver samples are frozen or fixed and culture medium collected for analysis.
6. Application of the porcine punch-excised intestinal explants to analyze intestinal permeability. Some mycotoxins such as DON induce intestinal barrier dysfunction and increase intestinal permeability in vivo and in vitro [9, 17]. Jejunal explants are a good alternative model to demonstrate the deleterious effects of DON on intestinal paracellular and transcellular permeability and to assess the protective effect of probiotic strains [18]. Jejunal explants, placed in Ussing chamber device, are bathed on each side with oxygenated KBH thermostatted at 37 °C. After equilibration (1 h), they are treated or not on the apical side with 10 μM DON for 3 h. Fluorescein sodium salt (FSS, 376 Da) and horseradish peroxidase (HRP, 44 kDa) are added in the apical compartment and measured in the basal compartment as markers of paracellular and transcellular permeability respectively.
7. Application of the porcine ex vivo models to analyze gene expression. Ex vivo models can also be used to assess the toxicity of mycotoxins on intestine and liver by transcriptomic analyses. Their total RNA is extracted, quantified, and retrotranscribed as cDNA. Quantitative polymerase chain reaction is performed using primers specific from genes of interest. As

DON frequently co-occurs with its glucosylated form, the deoxynivalenol-3-β-d-glucoside (D3G), their toxicity was compared using pig jejunal explants exposed to 10 μM D3G or DON for 4 h. An overexpression of pro-inflammatory cytokine genes was observed after DON exposure, while their expression in D3G-treated explants was similar to controls [19]. Similarly, DON increases the expression of transcription factors and inflammatory cytokines genes of pig PCLS exposed for 4 h to 10 μM DON [20]. A genome-wide transcriptomic analysis using an Agilent microarray was used to compare the effects of DON and two closely related recently discovered type A trichothecenes NX and 3ANX on pig intestinal explants. These three toxins regulate the same gene networks and signaling pathways (proliferation, differentiation, apoptosis, inflammation) but with a different magnitude: NX > DON ≫ 3ANX [21].

8. Application of the porcine ex vivo models to analyze protein abundance. Ex vivo models can also be used to analyze the effect of mycotoxins on targeted or global proteins abundance. For Western blot technique, samples are lysed with appropriate buffer containing protease inhibitors to extract proteins. After quantification, proteins are loaded on polyacrylamide gel, migrated, and transferred onto a membrane before being hybridized with primary and secondary antibodies. This technique was used to demonstrate the phosphorylation of selected mitogen-activated protein kinases (MAPKs) by DON [22]. Immunolabeling of tissue section allows the assessment of the expression of proteins of interest in tissue sections of explants or slices. Once deparaffinized with xylene and dehydrated, a heat-mediated antigen retrieval step is carried out and endogenous peroxidase activity is blocked with a methanol–-H_2O_2 solution. Sections are then incubated with antibodies and chromogen and finally counterstained with hematoxylin. Using this approach, PCLS exposed to DON revealed the activation of caspase-3, a key effector in cell apoptosis [20]. Global proteome changes can be addressed from explants or slices. Proteins were extracted in a buffer containing protease- and phosphatase inhibitors and then analyzed using gel-based (two-dimensional difference gel electrophoresis or 2DE-DIGE) or gel-free (nano-liquid chromatography–tandem mass spectrometry or nLC-MS/MS). The use of global proteomic analysis enabled a better understanding of the intestinal toxicity of the mycoestrogen ZEN and connected the estrogen activity of this mycotoxin with its intestinal toxicity [23].

9. Application of the porcine ex vivo models to analyze metabolites abundance. Ex vivo models can also be used to analyze metabolites abundance. Jejunal explants exposed to DON

during 4 h are extracted in methanol/dichloromethane/water and the 1H nuclear magnetic resonance spectra are generated using a Bruker spectrometer. The statistical integration of metabolomic and transcriptomic data increased the number of discriminant metabolites by a factor of 10, confirming the interest of the integration of both omics data [24].

10. Application of the porcine ex vivo models to analyze secretome. Proteomic analysis of leakage and tissue-specific secreted proteins (secretome) helps identify potential biomarkers. These proteins, typically functional within cells, may be released into plasma due to specific trafficking, signaling, or cellular injury. These may be direct indicators of liver damage, such as lactate dehydrogenase, aspartate or alanine aminotransferase [8], or released proteins, revealing the specific effects of treatments on liver tissue. For example, the comparison of the composition of the culture medium (secretome) of jejunal explants exposed to NX and DON is associated with a decreased viability and cell destruction. In addition, NX toxicity is associated with mitochondrial protein enrichment of the secretome, which may characterize its toxicity [25].
11. Application of the porcine PCLS to analyze mycotoxins biotransformation. PCLS mimic in vivo conditions and can metabolize xenobiotics. They can be used to identify metabolites of mycotoxin. Intestinal explants and PCLS were incubated with versicolorin A (VerA), a toxic precursor of AFB1, and metabolites were extracted and analyzed. Six metabolites were found, indicating liver glucuronidation and sulfation pathways. A metabolite suggesting VerA's bioactivation to a mutagenic form, potentially causing genotoxic effects, was also identified [26].
12. Application of the porcine ex vivo models to analyze histomorphology. One of the first readout to analyze intestinal or liver samples is the histomorphological analysis. Explants or slices are fixed with formaldehyde, dehydrated, and paraffin-embedded, and 5 μm sections are cut and stained with hematoxylin and eosin. A morphological lesion score, considering the extent of each lesion (according to the intensity or the frequency observed) and the severity factor, is then calculated and adapted for each scientific project (Table 2). This readout has been used thoroughly to study the effect of various mycotoxins [19, 20, 27–30].

Acknowledgments

This work was supported by the ANR grants "Precursors" (ANR-23-CE34-017) and "GenoMyc" (ANR-22-CE34-0022).

Table 2
Criteria considered for the histomorphological score

	Criteria	Severity factor	Maximum total score
Intestine	Cell vacuolation	1	54
	Debris	1	
	Enterocytes morphology	2	
	Loss of microvilli	2	
	Villi atrophy	2	
	Villi fusion	2	
	Edema	2	
	Necrosis	3	
	Villi number	3	
Liver	Trabecular disorganization	1	33
	Inflammatory infiltrate	1	
	Cytoplasmic vacuolation	1	
	Nuclear vacuolation	1	
	Megalocytosis	2	
	Apoptosis	2	
	Necrosis	3	

Maximum total score is obtained by the multiplication of the maximum intensity level of the lesions by the maximum severity factor for each organ

References

1. Cortez BRS, Guedes RMC (2023) A review on the evolution of methods for intestinal in vitro organ culture and its application in veterinary science. Vet World 16:347–356
2. Maresca M, Pinton P, Ajandouz EH, Menard S, Ferrier L, Oswald IP (2021) Overview and comparison of intestinal organotypic models, intestinal cells, and intestinal explants used for toxicity studies. Curr Top Microbiol Immunol 430:247–264
3. Dewyse L, De Smet V, Verhulst S, Eysackers N, Kunda R, Messaoudi N, Reynaert H, van Grunsven LA (2022) Improved precision-cut liver slice cultures for testing drug-induced liver fibrosis. Front Med 9:862185
4. de Graaf IA, Olinga P, de Jager MH, Merema MT, de Kanter R, van de Kerkhof EG, Groothuis GM (2010) Preparation and incubation of precision-cut liver and intestinal slices for application in drug metabolism and toxicity studies. Nat Protoc 5:1540–1551
5. Li M, de Graaf IA, Groothuis GM (2016) Precision-cut intestinal slices: alternative model for drug transport, metabolism, and toxicology research. Expert Opin Drug Metab Toxicol 12:175–190
6. Kolf-Clauw M, Castellote J, Joly B, Bourges-Abella N, Raymond-Letron I, Pinton P, Oswald IP (2009) Development of a pig jejunal explant culture for studying the gastrointestinal toxicity of the mycotoxin deoxynivalenol: histopathological analysis. Toxicol In Vitro 23:1580–1584
7. de Kanter R, Olinga P, Hof I, de Jager M, Verwillegen WA, Slooff MJ, Koster HJ, Meijer DK, Groothuis GM (1998) A rapid and simple method for cryopreservation of human liver slices. Xenobiotica 28:225–234
8. Dewyse L, Reynaert H, van Grunsven LA (2021) Best practices and Progress in precision-cut liver slice cultures. Int J Mol Sci 22:7137
9. Payros D, Garofalo M, Pierron A, Soler-Vasco-L, Al-Ayoubi C, Maruo VM, Alassane-Kpembi I, Pinton P, Oswald IP (2021) Mycotoxins in human food: a challenge for research. Cah Nutr Diet 56:170–183
10. Stoev SD (2024) Food security and foodborne mycotoxicoses-what should be the adequate risk assessment and regulation? Microorganisms 12:580
11. Alassane-Kpembi I, Puel O, Pinton P, Cossalter AM, Chou TC, Oswald IP (2017) Co-exposure to low doses of the food contaminants deoxynivalenol and nivalenol has a

synergistic inflammatory effect on intestinal explants. Arch Toxicol 91:2677–2687
12. Helke K, Nelson K, Sargeant A (2022) Animal models in toxicologic research: pig. In: Haschek and Rousseaux's (ed) Handbook of toxicologic pathology, 4th edn. Academic, pp 751–776
13. Krumdieck CL, dos Santos JE, Ho KJ (1980) A new instrument for the rapid preparation of tissue slices. Anal Biochem 104:118–123
14. Godoy P, Hewitt NJ, Albrecht U et al (2013) Recent advances in 2D and 3D in vitro systems using primary hepatocytes, alternative hepatocyte sources and non-parenchymal liver cells and their use in investigating mechanisms of hepatotoxicity, cell signaling and ADME. Arch Toxicol 87:1315–1530
15. Szalowska E, Stoopen G, Rijk JC, Wang S, Hendriksen PJ, Groot MJ, Ossenkoppele J, Peijnenburg AA (2013) Effect of oxygen concentration and selected protocol factors on viability and gene expression of mouse liver slices. Toxicol In Vitro 27:1513–1524
16. Obatomi DK, Brant S, Anthonypillai V, Early DA, Bach PH (1998) Optimizing preincubation conditions for precision-cut rat kidney and liver tissue slices: effect of culture media and antioxidants. Toxicol In Vitro 12:725–737
17. Pinton P, Tsybulskyy D, Lucioli J, Laffitte J, Callu P, Lyazhri F, Grosjean F, Bracarense AP, Kolf-Clauw M, Oswald IP (2012) Toxicity of deoxynivalenol and its acetylated derivatives on the intestine: differential effects on morphology, barrier function, tight junction proteins, and mitogen-activated protein kinases. Toxicol Sci 130:180–190
18. García GR, Payros D, Pinton P, Dogi CA, Laffitte J, Neves M, González Pereyra ML, Cavaglieri LR, Oswald IP (2018) Intestinal toxicity of deoxynivalenol is limited by *Lactobacillus rhamnosus* RC007 in pig jejunum explants. Arch Toxicol 92:983–993
19. Pierron A, Mimoun S, Murate LS, Loiseau N, Lippi Y, Bracarense AP, Liaubet L, Schatzmayr G, Berthiller F, Moll WD, Oswald IP (2016) Intestinal toxicity of the masked mycotoxin deoxynivalenol-3-β-D-glucoside. Arch Toxicol 90:2037–2046
20. Hasuda AL, Person E, Khoshal AK, Bruel S, Puel S, Oswald IP, Bracarense APFRL, Pinton P (2022) Deoxynivalenol induces apoptosis and inflammation in the liver: analysis using precision-cut liver slices. Food Chem Toxicol 163:112930
21. Pierron A, Neves M, Puel S, Lippi Y, Soler L, Miller JD, Oswald IP (2022) Intestinal toxicity of the new type A trichothecenes, NX and 3ANX. Chemosphere 288:132415
22. Lucioli J, Pinton P, Callu P, Laffitte J, Grosjean F, Kolf-Clauw M, Oswald IP, Bracarense AP (2013) The food contaminant deoxynivalenol activates the mitogen activated protein kinases in the intestine: interest of ex vivo models as an alternative to in vivo experiments. Toxicon 66:31–36
23. Soler L, Stella A, Seva J, Pallarés FJ, Lahjouji T, Burlet-Schiltz O, Oswald IP (2020) Proteome changes induced by a short, non-cytotoxic exposure to the mycoestrogen zearalenone in the pig intestine. J Proteome 224:103842
24. Tremblay-Franco M, Canlet C, Pinton P, Lippi Y, Gautier R, Naylies C, Neves M, Oswald IP, Debrauwer L, Alassane-Kpembi I (2021) Statistical integration of 'omics' data increases biological knowledge extracted from metabolomics data: application to intestinal exposure to the mycotoxin Deoxynivalenol. Metabolites 11:407
25. Soler L, Miller I, Terciolo C, Hummel K, Nöbauer K, Neves M, Oswald IP (2022) Exposure of intestinal explants to NX, but not to DON, enriches the secretome in mitochondrial proteins. Arch Toxicol 96:2609–2619
26. Al-Ayoubi C, Oules J, Person E, Bruel S, Bouville A, Pinton P, Oswald IP, Jamin EL, Puel O, Soler L (2022) Metabolism of versicolorin A, a genotoxic precursor of aflatoxin B1: characterization of metabolites using in vitro production of standards. Food Chem Toxicol 167:113272
27. Bracarense AP, Lucioli J, Grenier B, Drociunas Pacheco G, Moll WD, Schatzmayr G, Oswald IP (2012) Chronic ingestion of deoxynivalenol and fumonisin, alone or in interaction, induces morphological and immunological changes in the intestine of piglets. Br J Nutr 107:1776–1786
28. Terciolo C, Bracarense AP, Souto PCMC, Cossalter AM, Dopavogui L, Loiseau N, Oliveira CAF, Pinton P, Oswald IP (2019) Fumonisins at doses below EU regulatory limits induce histological alterations in piglets. Toxins 11: 548
29. Hasuda AL, Person E, Khoshal A, Bruel S, Puel S, Oswald IP, Bracarense APFRL, Pinton P (2023) Emerging mycotoxins induce hepatotoxicity in pigs' precision-cut liver slices and HepG2 cells. Toxicon 231:107195
30. Marquis V, Schulthess J, Oswald IP, Payros D (2023) Efficacité d'une fraction de levure enrichie en β-glucanes sur la réduction des effets toxiques induits par le déoxynivalenol *in vitro* et *ex vivo*. Journées Rech Porcine 55:383–388

Part III

Exposure Assessment

Chapter 15

Total Diet Study: An Example for Application of a Harmonized Public Health Methodology for Assessing Dietary Exposure to Chemical Contaminants

Elsa Vasco, Luísa Oliveira, and M. Graça Dias

Abstract

Total Diet Study (TDS) is a cost-efficient method recommended by WHO to estimate long-term exposure to chemicals in the diet. Based on national consumption data, commonly eaten foods are selected, prepared as consumed, pooled, and analyzed for contaminants. TDS provides average contamination levels but is not suitable for detecting short-term or isolated risks. Standardized methods, supported by projects like TDS-Exposure and EFSA guidelines, enhance data quality and global comparability. This harmonized TDS includes two main sampling phases: planning and fieldwork, followed by phases of chemical analysis, of the TDS samples and subsequent exposure assessment and risk characterization. In planning, core foods are chosen using national consumption data, categorized with the FoodEx2 system, and grouped into TDS samples. Sampling considers origin, market share, season, and cooking habits. Fieldwork covers food collection, preparation according to local cooking practices, and pooling of subsamples for analysis. Chemical analysis uses validated methods (e.g., ICP-MS, GC-MS, LC-MS/MS) and should use speciation for contaminants like arsenic and mercury. Exposure assessment combines contamination and consumption data using deterministic or probabilistic models and stratifies results by demographic groups. Risk characterization compares exposure estimates with health-based guidance values (HBGVs), using tools like tolerable intakes (TDI/TWI) and the Margin of Exposure (MOE), while accounting for variability and uncertainty through sensitivity analysis and Monte Carlo simulations. Results help identify major dietary sources of contaminants and support public health decisions.

Key words TDS food list, TDS samples, Food sampling methodology, Dietary exposure assessment

1 Introduction

Total Diet Study (TDS) is a cost-effective public health tool recommended by the World Health organization (WHO) to assess long-term population exposure to beneficial and harmful dietary chemicals [1]. The primary goal is to establish baseline data on chemical exposure levels and trends of a specific country's population based in foods as consumed by this population. TDS is mainly used to assess population chronic exposure to widely distributed

Paula Alvito and Helena Oliveira (eds.), *Food Contaminants Toxicity*, Methods and Protocols in Food Science,
https://doi.org/10.1007/978-1-0716-5096-7_15,

dietary chemicals. It is not suitable for acute exposure assessment or detecting high contamination in specific foods, as it is based on pooled samples analysis that only can provide mean concentration values [2].

TDS involves selecting, purchasing, and analyzing commonly consumed foods based on national food consumption data. These foods are grouped into a representative list, prepared as typically consumed, and pooled into samples for chemical analysis. Exposure levels are then estimated by combining analytical results with food consumption data. One of the main advantages of TDS is its cost efficiency, achieved by pooling samples before conducting analysis.

Harmonizing TDS methods across countries improves study comparability, efficiency, and accuracy in dietary exposure assessments. WHO published the first guidelines for the study of dietary intakes in 1985 [3], and in 2010, a Working Group from European Food Safety Authority (EFSA), Food and Agriculture Organization of the United nations (FAO) and WHO, the "Working Group of experts on TDS" developed a standardized approach, resulting in a comprehensive guidance document [1]. The European Commission later supported this effort by funding the Total Diet Study Exposure (TDS-Exposure) project (2012–2016), by the European Union's Seventh Framework Programme for research, technological development, and demonstration (Grant Agreement no. 289108).

TDS-Exposure aimed to harmonize TDS methodologies by ensuring consistency in food sampling, analytical procedures, exposure assessment modeling, and the prioritization of foods and chemicals [4].

Based on Portugal's participation in the TDS-Exposure project, this chapter describes a harmonized European Total Diet Study (TDS) methodology as an effective public health tool for assessing dietary exposure to chemical contaminants, enabling comparability, efficiency, and accuracy across studies.

2 Materials

1. To initiate a TDS, it is recommended to base the study on the results of a current food consumption survey conducted on the target population. This survey provides the necessary data to identify the most commonly consumed foods and to establish a representative food list for the study. So, it involves the selection of a list of core foods based on food consumption survey data collected at the individual level.
2. To ensure standardized classification and facilitate data comparability in this harmonized TDS, the FoodEx2 food classification and description system is used for coding and categorizing

foods. This system, developed by EFSA, allows for a harmonized approach to food classification, improving the accuracy and consistency of dietary exposure assessments.

3. The core foods, mapped to FoodEx2, are pooled into TDS samples representing the overall diet of the studied population. These selected foods are collected according to a defined sampling plan, prepared as consumed, and pooled before chemical analysis. The resulting analytical data is then combined with food consumption data to estimate dietary exposure to the targeted chemical substances. Aiming at comparable results across countries using a harmonized food classification system such as FoodEx2 is essential. However, dependent on the region, also other food classification systems might be appropriate.
4. The TDS follows a harmonized approach, implemented in two main sampling phases, planning and fieldwork [5], followed by chemical analysis of the TDS samples and subsequent exposure assessment and risk characterization.

2.1 Sampling Planning Phase

1. The TDS planning phase establishes a core food list mapped to FoodEx2 for pooling into samples. These samples represent the population diet under study, derived from individual food consumption data. A sampling plan is then designed based on the sample list, consumption data, market share, and local cooking habits.

2.1.1 Food Consumption Survey

1. The food consumption survey includes dietary consumption data that accurately represents the population under study.

   Participants are selected based on sex, age, and area of residence to ensure their representation of the target population.

   The survey should cover various dimensions, including socio-demographic, anthropometric, and food consumption data.

   Within the European Framework, EFSA recommended at least two non-consecutive 24-h recalls for data collection of food consumption data [6], but it is noted that this recommendation is not derived for specific TDS purposes.

2.1.2 Definition of the TDS Food List

1. Cleansing food consumption data.

   To process food consumption data in this TDS, individuals without recorded body weight are excluded from the original dataset. The remaining individuals are categorized into groups based on age and sex. To ensure harmonization and comparability with other European Total Diet Studies (TDS), the age group definitions follow the European Food Safety Authority

Input data from the table AVG FCD									Decision	Suggested TDS Food List for PT						
(sorted by this field)					Data from the national FCDB - raw, edible portion Average consumption (g/kg b.w./day)				Decision to include in TDS food list*	TDS Food list Name of composites are equal to FoodEx2 EN name			FCDB intended for next exposure assessment Average consumption (g/kg b.w./day)**			
					years	years	years	years					years	years	years	years
FoodEx2 Hierarchy code	FoodEx2 uniq code	FoodEx2ENname	National uniq code	National food name	sex	sex	sex	sex		FoodEx2 ENname	FoodEx2 uniq code	FoodEx2 Hierarchy code	sex	sex	sex	sex
				Total consumption	0	0	0	0					0	0	0	0

*Decision to include in TDS food list
1 Yes, as part of 90% of average consumption
2 Yes, as expected major contributor to exposure
3 Yes, other reason (see notes)
4 Combined into one item
N No, as not part of 90% of average consumption, nor expected major contributor

**In combined items the value presents sum of original average consumptions

Fig. 1 "TDS sample definition" template

(EFSA) guidelines for national food consumption data collection in pan-European dietary surveys [7].

The initial list of consumed foods is standardized, ensuring consistent spelling, resolving food name synonyms, and correcting typographical errors. Portion sizes are normalized across different measurement units, such as mass, volume, household measures, and pieces, so that all food items are consistently expressed in gram (g) or milliliter (mL).

Finally, the total daily consumption of each food item is calculated for each individual.

2. Food classification.

   For food classification, all food items are mapped to food groups. As described above, within Europe, 20 food groups at level one of the exposure hierarchy of the FoodEx2 food classification and description system [8] were used. Each individual food item is assigned a FoodEx2 hierarchy code, a FoodEx2 unique code, a FoodEx2 English name, a national unique code, and a national food name. This classification is then recorded in the "TDS sample definition" template (Fig. 1).

   The initial list of foods consumed by the selected population groups is aggregated into higher hierarchical food groups when necessary, particularly for highly detailed entries, with a recommended target of around 1000 foods or fewer, while ensuring full reversibility if needed [9].

3. Average food consumption.

   The consumption of each food item is calculated at the individual level, expressed in gram (g) or milliliter (mL) per kilogram of body weight per day. The average consumption of each food item is then determined separately for consumers and for the entire population, including both consumers and non-consumers. These calculations are performed across different age and sex groups, using the lowest viable food categorization level to distinguish as many foods as possible based on their codes. All results are recorded in the "TDS sample definition" template (Fig. 1).

4. TDS food list.

   Foods are selected for inclusion in the TDS food list based on two key criteria. The first criterion considers the most consumed foods, which contribute to, at least, 90% of overall food intake, including both consumers and non-consumers. The second criterion focuses on major contributors to exposure, including foods with particularly high levels of a given substance or those heavily consumed by specific sub-populations but not meeting the first criterion. These selection criteria are applied separately to the highest food aggregation level, which was in this harmonized TDS 20 FoodEx2 food groups at level one, ranking food items in descending order of consumption, excluding tap water, for each age and sex population group. The final selections are recorded in the "TDS sample definition" template (Fig. 1).

2.1.3 Pooling of TDS Food list Items into TDS Samples

1. The individual food items from the TDS food list are combined to define TDS samples using the "TDS Sample Definition" template (Fig. 1), applying one of two pooling methods—the Mixed Food Approach that combines various foods within the same food group (e.g., different species of fruit) or the Individual Food Approach that minimizes the extent of pooling by grouping similar items, such as different varieties of a single fruit species or various brands of the same product. The Individual Food Approach should be prioritized whenever possible. Separate commodities that are expected to be highly contaminated or consumed in large quantities should be kept separate to avoid dilution effects. As a starting point, it was expected that foods in the country-specific TDS food lists would be pooled into 100–300 TDS samples [9].
2. In this example of a harmonized TDS, each TDS sample consists of 12 pooled subsamples, selected based on statistical criteria for a pilot study [10], with the number of subsamples for each food item reflecting its relative consumption importance. Other TDS also could use more or less than 12 subsamples depending on the expected variability and available budgets.
3. For some samples, regional and seasonal variations should be considered. Regional TDS samples are included to capture variations in chemical levels due to local contamination or environmental factors. Unlike national samples, which are collected regardless of region, regional samples should be sourced across regions. Seasonality is considered for foods affected by supply origin or climate.
4. The list of TDS samples and their corresponding food items is then transferred to the "Food shopping list" template (Fig. 2) to support subsequent procedures.
5. To identify the specific varieties or brands of food items to collect, a combination of market share data sources is used,

TDS sample

Box 1 Sample characterization and number of subsamples (in national languages)

No.	TDS sample name:	Foodex2 code:		Total N subsamples:
1				12
	Items from the TDS food list	Characteristics (if needed):	Considered consump.	N of subsamples:
		Total	0.00	#DIV/0!

Box 2 Food items selection criteria (only in EN)

Nat/Reg:	Seasonal changes in TDS sample composition	National criteria for choosing of subsamples:	Place of purchase (type of store):

Box 3 Shopping list (in national languages)

	Food items purchased:	National criteria			Place of purchase	Purchased quantity and other details	
No.	Subsample	Type	Brand	Package		Min. (ml)	Notes
1							
2							
3							
4							
5							
6							
7							
8							
9							
10							
11							
12							

Fig. 2 "Food shopping list" template

Sampling plan - TDS EXPOSURE - pilot study
Campaign x - Season, year (month, month, month)

TDS sample number	FoodEx2 Code	TDS sample name	National (N) /Regional (R)	Week 1 when where	Week 2 when where	Week 3 when where	Week 4 when where	Week 5 when where
				Date 1 Place 1 Date 2 Place 2 Date 3 Place 3	Date 4 Place 4 Date 5 Place 5 Date 6 Place 6	Date 7 Place 7 Date 8 Place 8 Date 9 Place 9	Date 10 Place 10 Date 11 Place 11 Date 12 Place 12	

Fig. 3 "Food collection plan" template

including market research firms, national statistical offices, additional studies, and online resources. When such data are unavailable, brands or varieties are selected based on visibility and available quantity on store shelves. The quantity of each subsample and its ingredients is calculated, considering factors such as wastage (e.g., bones, shells, skin) during food preparation, as well as cooking-related gains (e.g., water absorption in rice and pasta) and losses (e.g., water or fat reduction).

2.1.4 Sampling Plan

1. When defining the sampling plan, key factors are considered, including food production and distribution regions, places of purchase, seasonal and varietal differences, brand variations, and potential contamination levels. Food collection regions are established based on these factors, with places of purchase and purchase proportions determined using market share data. Detailed information about each food item are recorded to support food collection planning, as well as the definition of food preparation and cooking methods.
2. Food collection is organized through multiple shopping campaigns conducted across different seasons, following the "Food collection plan" template (Fig. 3). To ensure practical

Campaign: Purchase date:

Place of purchase:

Purchase team	Purchasing managers	Chauffeur

Codes for corrective actions
1. Replacement with an item of the same range
2. Replacement with an item of a different range
3. Discontinued item
4. Other. Please specify.

Sample code	Subsample	Characteristic	Brand/Homemade	Amount	Unit	kind of storage	Preparation type	Ingredients	Corrective actions	Notes

Fig. 4 "Shop food list" template

efficiency, the national TDS sample collection is distributed across different campaigns. Specific details—such as campaign number, date, location, and purchase addresses—are documented in the "Food collection plan" template for each TDS sample.

3. A "Shop food list" template (Fig. 4) is developed based on the "Food collection plan" and used at purchase locations. This list includes:
    1. general information (campaign number, date, place, address, and collection team);
    2. subsample details (code, name, brand, quantity to be purchased);
    3. space for recording additional data.
4. Food packaging, labeling, transportation, and storage conditions are defined according to the characteristics of each food item. These conditions, along with other relevant details, are recorded in the "Food collection report" template, which is a database, which also includes the following information:
    1. TDS sample number and name;
    2. FoodEx2 code;
    3. subsample code and name;
    4. campaign number;
    5. purchase date;
    6. shop details;
    7. subsample characteristics (brand, producer, country of origin, lot number, shelf life, quantity purchased, and price);
    8. purchasing team details;
    9. cooking conditions;
    10. corrective actions and remarks.

2.2 Sampling Fieldwork Phase

1. During sampling fieldwork phase, food items are collected. TDS subsamples are prepared as consumed, pooled, and stored for laboratory analysis.

2.2.1 Organization of Food Collection and Sample Preparation

1. All necessary materials and equipment for food collection, transport, and preparation, including kitchen laboratory tools, need to be gathered or acquired. The kitchen laboratory should be set up, especially storage spaces as well as places for food preparation, taking into consideration the quantity and nature of the collected food items, as well as the number of samples to be homogenized and pooled. Appropriate storage conditions need to be ensured, including freezers (≤ -18 °C), refrigerators (4 ± 2 °C) and room temperature storage (≤ 25 °C).
2. Staff involved in food collection, transport, processing, and cooking should be trained using written instructions, assignment sheets, and checklists, with a strong emphasis on correctly handling any deviations from the established plan.
3. A standardized kitchen food preparation protocol needs to be developed to align with typical household cooking practices. It should be based on national culinary guidelines and/or instructions provided on food packaging. The protocol needs to be designed to reflect real consumer behavior, detailing appropriate cooking methods—such as boiling, grilling, braising, roasting, baking, frying, stewing, and microwaving—and specifying cooking times tailored to each method.

2.2.2 Food Collection

1. In this example of a harmonized TDS, shops are visited according to the "Food collection plan" and the "Shop food list." If certain food items are unavailable at the predefined locations, they are sourced from nearby shops within the same area. To maintain consistency, samples are primarily collected preferably on the same day of the week throughout all campaigns.
2. At least two individuals are required to conduct the food collection during each campaign. Samples are labeled and transported in appropriate containers, ensuring that required storage temperatures are maintained throughout the process. Upon arrival at the storage or laboratory facility, samples are stored under the correct temperature conditions.
3. Any additional information not provided on the packaging, such as country of origin, producer, or fruit variety for non-prepacked items, is recorded in the "Shop food list". Any deviations from the planned procedure that occur onsite are carefully documented.

2.2.3 Sample Reception

1. Upon arrival at the kitchen laboratory, each subsample is checked for compliance, and relevant information is recorded in the "Food collection report." All subsamples are labeled and photographed alongside a measurement scale in centimeter. For packaged subsamples, additional photographs are taken to ensure clear visibility of the product and brand

identification, ingredient list, nutritional information, shelf life, and barcode.

2.2.4 Sample Treatment and Storage

1. Standard culinary treatments are applied using namely, electric ovens, gas burners, and grills. Cooking is performed with stainless steel pans, non-stick aluminum frying pans, and utensils made of stainless steel, wood, plastic, porcelain, or glass, in accordance with national culinary practices. Inedible parts are removed either before or after cooking, depending on the type of food and consumer habits. Fruits and vegetables are washed, and foods are seasoned with olive oil (or the oil typically used in the country) and non-iodized/iodized salt, according to the use in the country. Tap water is used for cooking, again following typical country practices.
2. Equal amounts from 12 subsamples of each TDS sample are pooled and homogenized using a titanium-blade mixer, which helps prevent contamination from material migration. Depending on the analyte, its stabilization could be necessary before storage until analysis. The homogenized samples are then aliquoted into 100 mL HDPE (high density polyethylene) screw-cap containers, filled to capacity to minimize air exposure. Short-shelf-life subsamples are immediately frozen. When such subsamples from one TDS sample are collected at different times, they are stored in a 2 L HDPE container and frozen until the complete TDS sample set is collected.
3. All containers are labeled with waterproof stickers and ink, indicating the date of packaging, food group, and TDS sample number. Analytical aliquots, prepared according to specific analytical procedures, are frozen at ≤ -18 °C until laboratory analysis. At least one container per sample is retained in frozen storage until the completion of the study and publication of the results.

2.3 Chemical Analysis

1. In a TDS, the determination of contaminants requires the use of robust and reliable analytical methods, preferably those accredited under standards such as ISO/IEC 17025 [11]. At a minimum, all methods must be properly validated to ensure the accuracy, precision, and reliability of results.
2. Using the health-based guidance values (HBGVs) for each substance, together with the TDS sample list, the consumption levels, and the analytical detection limits, an assessment is performed. This assessment determines whether the current laboratory detection limits are sufficiently low to protect consumer health, or whether lower (more sensitive) analytical limits are required.

3. Commonly employed techniques include inductively coupled plasma mass spectrometry (ICP-MS) for the analysis of heavy metals, direct mercury analyzer (DMA) for mercury, gas chromatography–mass spectrometry (GC-MS) and liquid chromatography coupled to tandem mass spectrometry (LC-MS/MS) for the detection of pesticide residues and persistent organic pollutants, and high-performance liquid chromatography (HPLC) for the quantification of mycotoxins and nitrates, among others.
4. Given that different chemical species and forms of contaminants can vary significantly in their toxicity and impact on human health, speciation analysis is essential for accurate risk assessment. For instance, arsenic speciation using liquid chromatography-inductively coupled plasma mass spectrometry (HPLC-ICP-MS) allows for the differentiation between toxic and less harmful forms, providing more meaningful exposure assessments.

2.4 Exposure Assessment and Risk Characterization

1. Exposure assessment integrates food contamination data with food consumption data obtained from national dietary surveys. Individual consumption data, stratified by age, gender, and population subgroups, are used to estimate daily intake levels. Methods include:
   1. Deterministic Approach: Uses average contamination levels and mean or high-percentile consumption data to estimate exposure.
   2. Probabilistic Approach: Employs statistical distributions to model variability and uncertainty in food consumption and contaminant levels, providing a more comprehensive risk assessment.
   3. In between, due to limitations, semi-probabilistic approaches are sometimes used.
2. Exposure estimates are calculated for food as measured (TDS samples) and for food as eaten (TDS subsamples) and are expressed using statistical indicators such as mean, median, (P50), and 95th percentile (P95) to represent average and highly exposed populations. Exposure is expressed in relation to body weight (bw) (e.g., μg/kg bw/day or μg/kg bw/week) and is conducted for the overall population as well as segmented by age and gender groups.
3. Risk characterization involves comparing estimated exposure levels with HBGVs established by international food safety authorities (e.g., EFSA, WHO, FAO). Margin of exposure (MOE) approach may be used for contaminants without established tolerable intake levels (as the Benchmark Dose Limit (BMDL)), while for others, comparisons are made with

tolerable daily intake (TDI) or tolerable weekly intakes (TDI/TWI). The percentage of individuals exceeding these reference values is also estimated, providing insights into potential public health risks.

4. A comprehensive risk assessment also includes evaluating the uncertainty and variability in exposure estimates. Sensitivity analyses may be performed to assess the influence of different exposure scenarios and methodological assumptions. Additionally, probabilistic modeling techniques such as Monte Carlo simulations can be employed to refine risk characterization by generating distributions of exposure estimates rather than point estimates.
5. The contribution of different food groups to total exposure is estimated to identify major dietary sources of contaminants. This information helps prioritize risk management actions, such as targeted monitoring programs, regulatory measures, and public health interventions aimed at reducing exposure to potentially harmful substances.

3 Notes

TDS is a cost-effective approach for assessing dietary exposure to contaminants, as it maximizes resources by pooling multiple food items into composite samples while still providing valuable insights into population-wide exposure. This efficiency allows for broader monitoring of contaminants with relatively lower analytical costs, making TDS a practical and scalable tool for food safety authorities and public health decision-makers. By incorporating key recommendations such as those outlined by EFSA—including robust food list design, strategic inclusion of high-risk foods, and standardized classification systems—TDS studies can yield high-quality data that support informed risk assessment and regulatory decisions.

In this context, primary notes are:

1. Representative Food Sampling Is Essential: Ensure that selected food items reflect actual consumption patterns to provide meaningful exposure estimates [12].
2. Validated Analytical Methods Improve Data Reliability: Use accredited or at least validated methods to ensure accurate contaminant detection and quantification. Verify if limits of quantification are adequate.
3. Handling of Left-Censored Data Requires Standardized Approaches: Utilize internationally accepted substitution methods (upper bound (UB)/lower bound (LB)) to manage non-detected values appropriately [13].

4. Robust Exposure Assessment Models Enhance Accuracy: Combining deterministic and probabilistic approaches improves exposure estimation, capturing variability within the population.
5. Integration with Food Consumption Data Is Key: Link occurrence data with national dietary surveys using standardized classification systems to ensure consistency in exposure calculations.
6. Risk Characterization Should Consider Uncertainty: Monte Carlo simulations and sensitivity analyses can enhance the reliability of exposure estimates and risk conclusions.
7. Food Contribution Analysis Supports Risk Management: Identifying major dietary sources of contaminants helps in formulating targeted public health interventions and regulatory measures.
8. Continuous Review and Methodological Updates Are Necessary: Regular updates to methodologies, based on new scientific evidence and regulatory developments, are crucial for maintaining the relevance of TDS studies.
9. Following the Guideline Given by a Joint WHO, FAO, and EFSA Working Group Enhances Study Quality: Using EFSA's 90% threshold for constructing a TDS food list ensures comprehensive coverage of food consumption patterns [12].
10. Prioritizing Foods Susceptible to Contamination Is Critical: Including foods known to have high contaminant concentrations enhances exposure assessment accuracy.
11. Coding Consistency Facilitates Data Integration: A standardized food coding system (e.g., FoodEx2) ensures compatibility with international classification frameworks [12].
12. Consideration of Subpopulation Dietary Patterns Strengthens Assessments: Depending on the budget, lower levels of food aggregation enable to account for separately, for example, young children, vegetarians, and highly exposed groups.

References

1. European Food Safety Authority [EFSA], Food and Agriculture Organization of the United Nations [FAO] WHO [WHO] (2011) Towards a harmonised total diet study approach: a guidance document. EFSA J 9(11):1–66
2. Moy GG, Vannoort RW (2013) Total diet studies. Springer, New York, pp 1–550
3. WHO (1985) Guidelines for the study of dietary intakes of chemical contaminants. WHO Offset Publ 87:1–104
4. Papadopoulos A, Sioen I, Cubadda F, Ozer H, Oktay Basegmez HI, Turrini A et al (2015) TDS exposure project: application of the analytic hierarchy process for the prioritization of substances to be analyzed in a total diet study. Food Chem Toxicol [Internet] 76:46–53. Available from: https://linkinghub.elsevier.com/retrieve/pii/S0278691514004827
5. Vasco E, Dias MG, Oliveira L (2021) The first harmonised total diet study in Portugal: planning, sample collection and sample

preparation. Food Chem [Internet] 363 (June):130258. https://doi.org/10.1016/j.foodchem.2021.130258
6. European Food Safety Authority [EFSA] (2009) General principles for the collection of national food consumption data in the view of a pan-European dietary survey. EFSA J 7(12): 1435. (1–51)
7. European Food Safety Authority [EFSA] (2011) Use of the EFSA comprehensive European food consumption database in exposure assessment. EFSA J 9(3):2097. (1–34)
8. European Food Safety Authority [EFSA] (2011) The food classification and description system FoodEx 2 (draft-revision 1). EFSA Support Publ 215:1–438
9. Dofkova M, Nurmi T, Berg K, Reykdal Ó, Gunnlaugsdóttir H, Vasco E et al (2016) Development of harmonised food and sample lists for total diet studies in five European countries. Food Addit Contam Part A [Internet] 33(6):933–944. https://doi.org/10.1080/19440049.2016.1189770
10. Julious SA (2005) Sample size of 12 per group rule of thumb for a pilot study. Pharm Stat 4: 287–291
11. European Committee For Standardization. EN ISO/IEC 17025:2005 (2005) General requirements for the competence of testing and calibration laboratories. Brussels
12. Akhandaf Y, Henauw S, Dofkova M, Ruprich J, Papadopoulos A, Sirot V et al (2015) Establishing a food list for a total diet study: how does food consumption of specific subpopulations need to be considered? Food Addit Contam Part A [Internet] 32(1):9–24. https://doi.org/10.1080/19440049.2014.984776
13. European Food Safety Authority [EFSA] (2010) Management of left-censored data in dietary exposure assessment of chemical substances. EFSA J 8(3):1557. (1–96)

Chapter 16

Food Contaminants Exposure Assessment Using Random Forest Model

Sofia Almeida Costa, Milton Severo, Catarina Carvalho, Carla Lopes, and Duarte Torres

Abstract

In toxicological studies, exposure assessment is a crucial step for determining exposure to food contaminants. Selecting an appropriate method is critical to exposure assessments. Here, we describe an alternative approach to assess exposure to food contaminants, accounting for individual variability. This methodological approach offers a more comprehensive and realistic perspective by capturing real-world exposure scenarios, thereby enhancing the relevance and applicability of the findings.

Key words Bisphenol A, Food contaminants, Exposure assessment, Intake, Urinary biomarkers

1 Introduction

Food contaminants can occur naturally in the food matrix due to environmental contamination, be non-intentionally formed during production, processing, treatment, transport, storage, preparation, and cooking, or migrate from packaging materials [1]. Exposure to these compounds may adversely affect human health and requires monitoring and control [2].

According to the Codex Alimentarius [3], within the risk analysis of food hazards, risk assessment is essential for applying scientific knowledge in establishing food safety standards, guidelines, and recommendations that support decision-making. The risk assessment approach uses available scientific data to evaluate the risk associated with exposure to food contaminants. It follows four steps: hazard identification, hazard characterization, exposure assessment, and risk characterization [3].

In addition to hazard identification and hazard characterization, exposure assessment and risk characterization are crucial steps for publishing rigorous, relevant, and transparent scientific

Paula Alvito and Helena Oliveira (eds.), *Food Contaminants Toxicity*, Methods and Protocols in Food Science, https://doi.org/10.1007/978-1-0716-5096-7_16,

opinions on public health concerns related to specific food contaminants.

The scope of the exposure assessment is to assess chronic exposure to food contaminants. Selecting an appropriate method is crucial for carrying out exposure assessment studies, as it depends on the purpose of the study, the nature of the contaminant, the available data, and the available resources [4]. As an indirect method, theoretical daily dietary exposure is the most commonly used approach that combines food consumption data with estimates of chemical concentration in foodstuffs.

Individual food consumption data can be collected through self-reported methods, namely, 24-h recalls or food diaries, while food contaminant concentration in food products can be obtained by laboratory analysis or gathered from the literature previously published [4, 5]. However, self-reported food consumption information has several limitations related to misreporting, recall bias, participation bias, and identification of all dietary sources of food contaminants [6–8]. Furthermore, depending on the hazard compound, dietary assessment methods may be insufficient to account for intra-individual variability in usual food intake [9] and, consequently, fail to capture the entire picture of food contaminants' exposure.

Using direct analytical methods, exposure to human food contaminants can be evaluated by measuring the concentration of the hazard compound or its metabolites in biological samples. Biomarkers, such as urinary metabolites or hemoglobin adducts, provide objective estimates and may be used to validate and improve indirect methods.

Thus, researchers must consider a broad set of factors and intake determinants to make informed decisions based on scientific evidence, ensuring an accurate integration of food consumption information with data on the occurrence of contaminant while evaluating a scenario representative of the real-life situation.

This chapter proposes applying a Random Forest statistical model to estimate daily exposure to bisphenol A, integrating data from self-reported methods and urinary metabolites as an alternative methodological approach.

2 Methods

This chapter will describe the Random Forest statistical model as an alternative methodological approach to estimate daily bisphenol A (BPA) exposure. For that, data from self-reported dietary assessment methods and the urinary concentration of BPA were used. This approach improves the predictions, widening the range of exposure estimates by accounting for the intra- and inter-individual variability, usually underestimated in the deterministic models.

2.1 Predictor Variables

Food diaries or 24-h recalls are examples of dietary assessment methods used to collect food consumption information needed for exposure assessment studies and, consequently, risk assessment studies. Considering the BPA characteristics and the occurrence of this contaminant in food [10], collecting dietary information regarding food items/groups and the associated packaging material is essential. In the Random Forest model, these will be the predictor variables.

In our approach, dietary information was collected or completed via face-to-face interviews by trained professionals, following the EU Menu methodology proposed by the European Food Safety Authority [11]. Food data was collected and processed using the validated "eAT24" software [12, 13]. The participants were asked to report, quantify, and describe all items consumed at each eating occasion, including beverages and dietary supplements.

The FoodEx2 classification system was implemented, allowing a complete description of all food items reported through several facets and corresponding descriptors [14]. Thus, it was possible to collect a large set of information about the participants' food intake, including the preparation-producing place, cooking process and extent of cooking, and food packaging materials (pottery, wood, cork, glass, paper, paperboard, vegetal paper, film, plastic, textiles, multilayer materials, metal, aluminum, and edible and other materials).

Finally, foods, beverages, and recipes were quantified through different methods available: weight (grams), volume (milliliters), food picture book (186 food photo series for foods/recipes (six portions each) and 11 photo series for household measures) [15], standard units, household measures, and default portions.

2.2 Outcome Variable

BPA is excreted primarily via urine [16], with a half-life of nearly six hours in the human body, making this a valid biomarker of short-term daily exposure [17, 18]. Considering the low contribution of other routes, the concentration of BPA in urine may represent dietary exposure to this contaminant for the general population [19–21]. Thus, the 24-h urine collection allows us to estimate individual BPA exposure over an entire day with high precision and accuracy. However, this method is demanding and costly for study participants, requiring substantial human and financial resources that are often limited or unavailable. Because of that, usually, in population-based studies, 24-h urine samples are collected only for a sub-sample of participants. Applying the Random Forest Model, daily BPA exposure for the remaining sample can be predicted.

BPA concentration levels in urine can be measured using a dispersive liquid-liquid microextraction followed by gas chromatography coupled to mass spectrometry, as previously described [22]. Total BPA excretion (ng/day) was calculated by multiplying the urinary BPA concentration (ng/L) and the urine volume (L/day). A Box-Cox transformation was performed to approximate the distribution of the total urinary BPA to the normal distribution curve, applying a calculated exponent of −0.2 according to our sample characteristics. Urinary BPA concentration measured from a sample (or a sub-sample) will be the outcome variable in the Random Forest model.

2.3 Development of the Random Forest

The Random Forest model was built using the *randomForest* function from the *randomForest* package in R [23] to model the relationship between predictor variables and the outcome. Random Forest combined the various outcome predictions from multiple independent regression trees to reach a single outcome result, considering the different predictor variables. This model requires only two parameters: the number of trees (ntree) and the number of randomly selected predictors per decision point (mtry). A small ntree may lead to high variance and compromise the model development. Different values should be tested (for example, 100, 200, 500, and 1000) to determine where the model's performance stops improving. Regarding mtry, a higher value may lead to less randomness but increases the correlation between trees, increasing the risk of overfitting. To assess the optimal mtry value, root mean square error (RMSE), mean absolute error (MAE), or the coefficient of determination (R-squared - R^2) can be used.

In the present study, 1000 ntree were generated to create the Random Forest, using food groups and the packaging material reported as predictors. Moreover, to handle the possible overfitting of the models, a repeated 2-times 5-fold cross-validation was performed. According to our data, the optimal selected mtry was two since it presented lower values for RMSE and MAE and a higher R^2 (Figs. 1, 2, and 3). Interpreting Figs. 1, 2, and 3, it becomes evident that increasing the number of randomly selected predictors does not improve model performance. In fact, higher numbers of randomly selected predictors were associated with increased RMSE and MAE, and a decline in the correlation between predicted and actual values.

In addition to aggregated predictions, the Random Forest model also provides a variable importance plot, which shows the percentage increase in mean squared error (%IncMSE) caused by randomly permuting each variable.

Figure 4 shows an example of a variable importance plot for predicting BPA daily exposure. Considering our example, the consumption of powder, condensed, and evaporated milk was the variable with higher importance for the estimation of BPA exposure (5.5%IncMSE), followed by the consumption of food items packaged in plastic (4.5%IncMSE) (Fig. 4). This indicates that excluding only these two variables from the model would lead to an

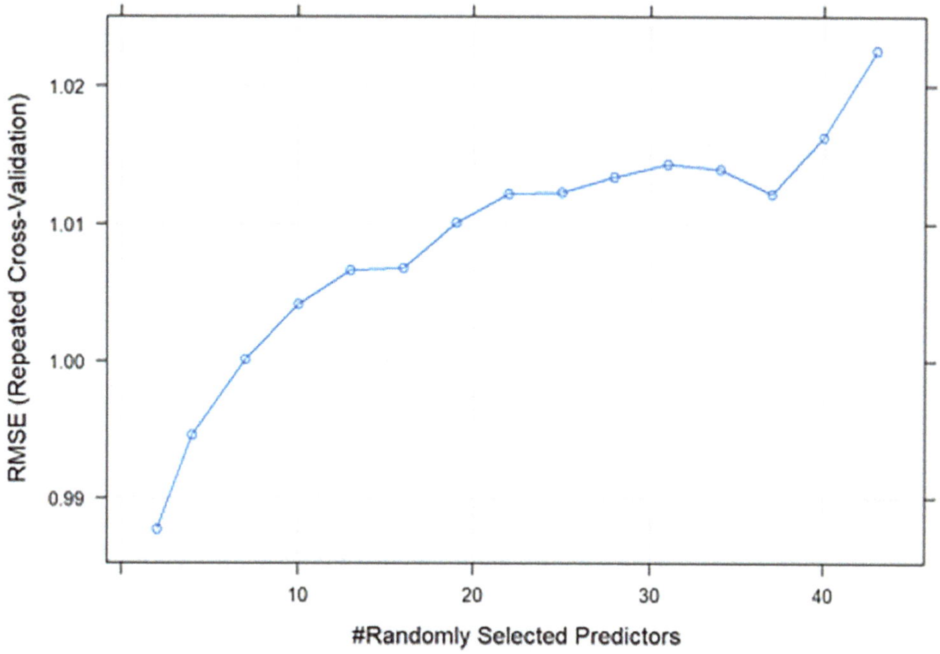

Fig. 1 Plot of fit-index parameter root mean squared error (RMSE) after cross-validation

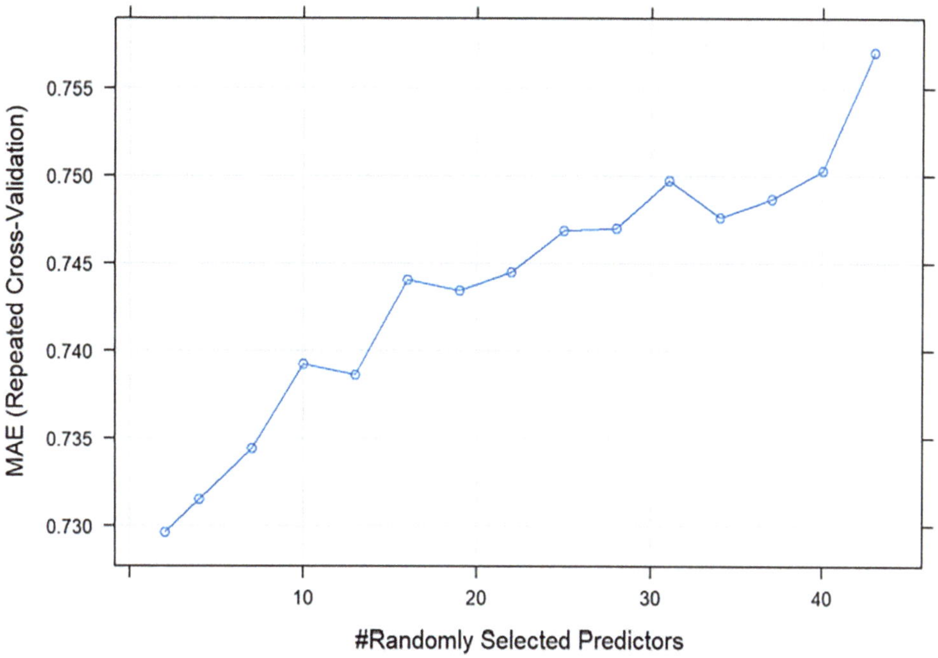

Fig. 2 Plot of fit-index parameter mean absolute error (MAE) after cross-validation

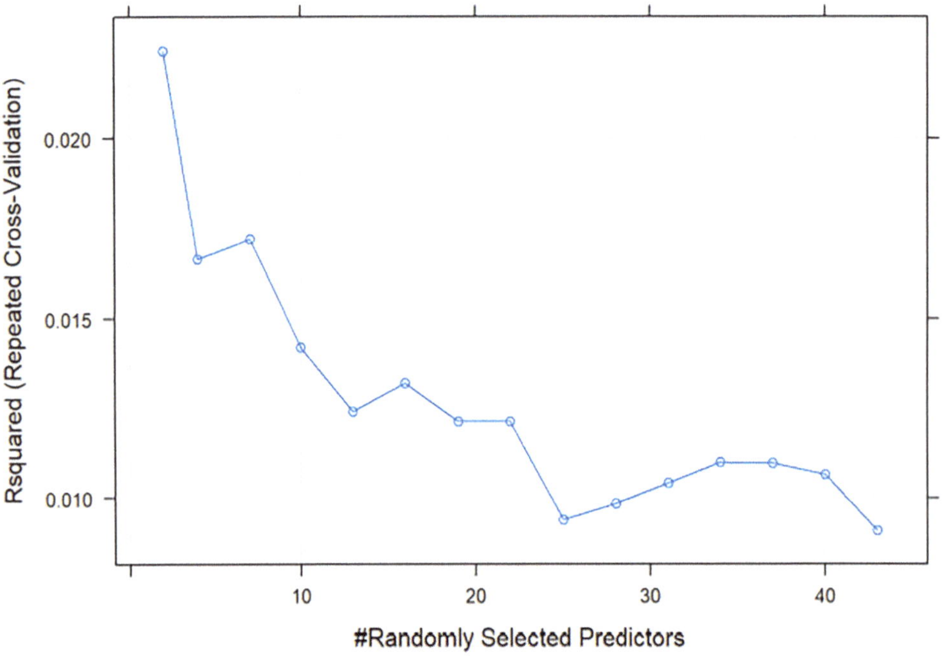

Fig. 3 Plot of fit-index parameter R-squared (R^2) after cross-validation

approximate 10% increase in the model's prediction error, highlighting their importance in maintaining predictive performance.

These results were reasonably expected since, in the European Union, BPA was authorized until 2025 to be used in the production of polycarbonate plastics and epoxy resins of food contact materials [24, 32]. Thus, BPA is frequently found in reusable beverage bottles, tableware, cookware, microwave ovenware, food containers, and reservoirs for water dispensers made with polycarbonate plastics [10, 19]. Moreover, cans, vats, tanks, and supply systems in contact with foods and drinks are coated with protective linings containing epoxy resins composed of BPA [10].

BPA can leach from food contact materials and migrate into foods and beverages. This process is influenced by storage conditions such as temperature, pH of the surrounding medium, water content, area, and contact time [25, 26]. Therefore, the most important source of BPA exposure for the general population is diet [10, 27], primarily through canned food and beverages [28, 29].

Then, using the Random Forest model generated on data from our sub-sample, daily BPA exposure was estimated for the remaining participants with missing urinary BPA concentration data but complete dietary information. This methodological approach incorporated intra- and inter-individual variability to develop a more precise and accurate model that better reflects real-world

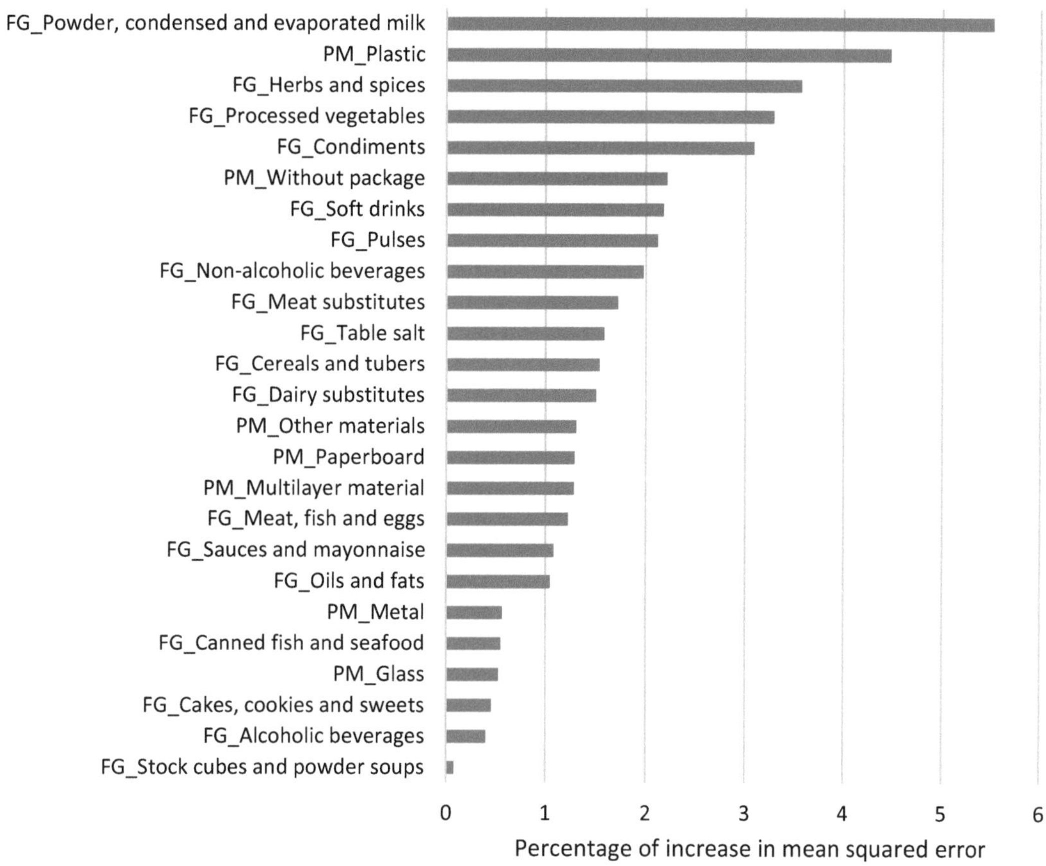

Fig. 4 Percentage of increase in mean squared error according to the Random Forest variable importance, computed for our sample. Abbreviations: FG - food group; PM - packaging material

exposure. An application example of the method can be found in Costa et al., 2023 [30].

Our findings suggested that random forest was the best methodology to estimate the daily exposure to BPA by comparing with the regression tree and food group attribution (deterministic approach, using only self-reported dietary information). According to the results, the random forest presented the greatest fit-index between urinary BPA and the estimated daily exposure, for both adolescents and adults, in full data analysis and after the cross-validation [30]. The deterministic approach proved to be less accurate in estimating daily BPA exposure, with median values in adolescents being five to six times higher than those measured in urine [30].

3 Notes

1. The choice of the method for collecting dietary information is crucial and depends on the contaminant under investigation. The method's sensitivity and specificity must be carefully considered to accurately capture intra- and inter-individual variability.
2. The collection of dietary information should be as comprehensive as possible, including detailed descriptions of food product characteristics essential for building models such as the Random Forest. These characteristics are usually specific to each food contaminant.
3. The dietary information collected and the BPA urinary measurement must correspond to the same day.
4. Selecting optimal values for ntree and mtry is crucial for achieving good performance in a Random Forest model.
5. To reduce possible overfitting of the model, a cross-validation can be applied.
6. Considering the half-life of BPA, 24-h urine samples are preferable to spot urine samples. However, in addition to the higher associated costs of direct methods, they are challenging to implement and have specific limitations [31].
7. Even with a limited number of urine samples, Random Forest predicts the daily BPA exposure for the remaining sample by considering the information of the predictor variables.
8. It is important to highlight that biomarkers account for all exposure sources, not only daily dietary exposure. The model presented in this chapter can only be applied to food contaminant exposure if other sources of contaminant exposure are considered negligible and the biomarker provides a reliable representation of dietary exposure.

References

1. Rather IA, Koh WY, Paek WK, Lim J (2017) The sources of chemical contaminants in food and their health implications. Front Pharmacol 8(NOV)
2. Severin I, Souton E, Dahbi L, Chagnon MC (2017) Use of bioassays to assess hazard of food contact material extracts: state of the art. Food Chem Toxicol [Internet] 105:429–447. Available from: https://www.scopus.com/inward/record.uri?eid=2-s2.0-85019138979&doi=10.1016%2Fj.fct.2017.04.046&partnerID=40&md5=68fa6277d77217f9c5c07ac43a67a219
3. Food and Agriculture Organization of the United Nations, World Health Organization (2023) Codex Alimentarius Commission Procedural Manual [Internet]. 28th ed. Rome. Available from: https://doi.org/10.4060/cc5042en
4. Lambe J (2002) The use of food consumption data in assessments of exposure to food chemicals including the application of probabilistic modelling. Proc Nutr Soc 61(1):11–18
5. de Fátima PM, Hogg T (2007) Exposure assessment of chemicals from packaging

materials in foods: a review. Trends Food Sci Technol 18:219–230
6. Brouwer-Brolsma EM, Brennan L, Drevon CA, Van Kranen H, Manach C, Dragsted LO et al (2017) Combining traditional dietary assessment methods with novel metabolomics techniques: present efforts by the food biomarker Alliance. Proc Nutr Soc 76(4):619–627
7. Food and Agriculture Organization of the United Nations. 2018. Dietary assessment: a resource guide to method selection and application in low resource settings [Internet]. Vitamin B-6 metabolism in pregnancy, lactation, and infancy. Rome. Available from: https://openknowledge.fao.org/server/api/core/bitstreams/3dc75cfc-9128-4f29-9d76-8d1f792078f0/content
8. Bingham SA, Bingham SA, Bingham AM, Bingham AM. The dietary assessment of individual: methods, accuracy, new techniques and recommendations. In 1987. Available from: https://api.semanticscholar.org/CorpusID:79511108
9. Thompson FE, Subar AF (2013) Dietary assessment methodology. In: Coulston AM, Boushey CJ, Ferruzzi MG (eds) Nutrition in the prevention and treatment of disease, 3rd edn. Academic, London
10. European Food Safety Authority (2023) Re-evaluation of the risks to public health related to the presence of bisphenol A (BPA) in foodstuffs. EFSA J [Internet] 21(4):e06857. https://doi.org/10.2903/j.efsa.2023.6857
11. European Food Safety Authority (2014) Guidance on the EU menu methodology. EFSA J 12(12)
12. Lopes C, Torres D, Oliveira A, Severo M, Guiomar S, Alarcão V et al (2018) National food, nutrition, and physical activity survey of the Portuguese general population (2015–2016): protocol for design and development. J Med Internet Res [Internet] 20(2): 1–11. Available from: www.ian-af.up.pt
13. Goios ACL, Severo M, Lloyd AJ, Magalhães VPL, Lopes C, Torres DPM (2020) Validation of a new software eAT24 used to assess dietary intake in the adult Portuguese population. Public Health Nutr 3:1–11
14. European Food Safety Authority (2011) Report on the development of a food classification and description system for exposure assessment and guidance on its implementation and use. EFSA J 9(12):1–84
15. Torres D, Faria N, Sousa N, Soares R, Amorim H, Guiomar S et al (2017) Manual Fotográfico de Quantificação de Alimentos [Internet]. Universidade do Porto, Available from: www.ian-af.up.pt
16. EU-RAR (European Union Risk Assessment Report) 2008. 4,4′-isopropylidenediphenol (Bisphenol-A). Human Health, Part 2. Publications Office of the European Union [Internet]. 2008; Available from: https://publications.jrc.ec.europa.eu/repository/handle/JRC59988
17. Stahlhut RW, Welshons WV, Swan SH (2009) Bisphenol a data in NHANES suggest longer than expected half-life, substantial nonfood exposure, or both. Environ Health Perspect 117(5):784–789
18. Thayer KA, Doerge DR, Hunt D, Schurman SH, Twaddle NC, Churchwell MI et al (2015) Pharmacokinetics of bisphenol A in humans following a single oral administration. Environ Int 83:107–115
19. Vandenberg LN, Hauser R, Marcus M, Olea N, Welshons WV (2007) Human exposure to bisphenol A (BPA). Reprod Toxicol 24(2): 139–177
20. Meslin M, Beausoleil C, Zeman FA, Antignac JP, Kolossa-Gehring M, Rousselle C et al (2022) Human biomonitoring guidance values (HBM-GVs) for Bisphenol S and assessment of the risk due to the exposure to Bisphenols A and S, in Europe. Available from: https://doi.org/10.3390/toxics10050228
21. Von Goetz N, Wormuth M, Scheringer M, Hungerbühler K (2010) Bisphenol A: how the most relevant exposure sources contribute to total consumer exposure. Risk Anal 30(3): 473–487
22. Cunha SC, Fernandes JO (2010) Quantification of free and total bisphenol A and bisphenol B in human urine by dispersive liquid-liquid microextraction (DLLME) and heart-cutting multidimensional gas chromatography-mass spectrometry (MD-GC/MS). Talanta [Internet] 83(1): 117–125. https://doi.org/10.1016/j.talanta.2010.08.048
23. Liaw A, Wiener M (2002) Classification and regression by random forest. R Journal [Internet] 2(3):18–22. Available from: https://cran.r-project.org/doc/Rnews/Rnews_2002-3.pdf
24. European Commission (2011) Commission Regulation (EU) No 10/2011 of 14 January 2011 on plastic materials and articles intended to come into contact with food. Off J Eur Union [Internet]. Available from: https://eur-lex.europa.eu/eli/reg/2011/10/oj
25. Noureddine El Moussawi S, Cladière M, Chébib H, Ouaini R, Camel V (2019) Empirical models to predict the effect of sterilisation

and storage on bisphenols migration from metallic can coatings into food simulants. Food Addit Contam Part A Chem Anal Control Expo Risk Assess 36(12):1937–1949

26. Noureddine El Moussawi S, Ouaini R, Matta J, Chébib H, Cladière M, Camel V (2019) Simultaneous migration of bisphenol compounds and trace metals in canned vegetable food. Food Chem [Internet] 288:228–238. Available from: https://www.sciencedirect.com/science/article/pii/S0308814619304480
27. Martínez MA, Rovira J, Prasad Sharma R, Nadal M, Schuhmacher M, Kumar V (2018) Comparing dietary and non-dietary source contribution of BPA and DEHP to prenatal exposure: a Catalonia (Spain) case study. Environ Res [Internet] 166:25–34. Available from: https://www.sciencedirect.com/science/article/pii/S0013935118302470
28. Chang WH, Liu SC, Chen HL, Lee CC (2019) Dietary intake of 4-nonylphenol and bisphenol A in Taiwanese population: integrated risk assessment based on probabilistic and sensitive approach. Environ Pollut (244):143–152
29. Liu J, Wattar N, Field CJ, Dinu I, Dewey D, Martin JW (2018) Exposure and dietary sources of bisphenol A (BPA) and BPA-alternatives among mothers in the APrON cohort study. Environ Int [Internet] 119(March):319–326. https://doi.org/10.1016/j.envint.2018.07.001
30. Costa SA, Severo M, Correia D, Carvalho C, Magalhães V, Vilela S et al (2023) Methodological approaches for the assessment of bisphenol A exposure. Food Res Int [Internet] 173:113251. Available from: https://www.sciencedirect.com/science/article/pii/S0963996923007962
31. International Life Sciences Institute (2002) Report series – exposure from food contact materials – summary report of a workshop held in October 2001 [Internet]. Ispra, Italy. Available from: https://ilsi.eu/wp-content/uploads/sites/3/2016/06/R2002Exp_Food.pdf
32. European Commission. Commission Regulation (EU) 2024/3190 of 19 December 2024 on the use of bisphenol A (BPA) and other bisphenols and bisphenol derivatives with harmonised classification for specific hazardous properties in certain materials and articles intended to come int. Official Journal of the European Union. 2024; Available from: https://data.europa.eu/eli/reg/2024/3190/oj

Index

A

Acid digestion 35–45, 150
Acrylamide (AA) 47–56
Adsorption 16, 141, 142, 146–148, 150–153
Africa 57, 81–99

B

Beef 26
Bisphenol A (BPA) 196–198, 200–202

C

Cell viability 104, 107–109, 119, 150, 163
Cereals 6, 7, 37, 48, 49, 55, 90
Chicken 26, 86, 168
Co-exposure 141–154
Contaminants 12, 37, 47, 48, 68, 69, 88, 89, 113–120, 141, 142, 147, 148, 167, 168, 181–192, 195–197, 202
Cultured cells 114
Cytotoxicity 103, 109, 129, 134, 158, 159, 161–163

D

Dietary exposure assessments 182, 183
Digested-micro and nanoplastics 157–163

E

Engineered nanomaterials (ENM) 141, 144–145
ENM–PTE interactions 141, 142, 147, 148, 152
Environmental pollutants 72
Enzyme-linked immunosorbent assay (ELISA) 81–99
Explants 167–175
Exposure assessment 182, 183, 190–192, 195–202
Ex vivo models 167–175

F

Fish 16, 19, 35, 37, 158
Food analysis 13, 49
Food contaminants 104, 114, 195–202
Food contamination 190
Food/feed contaminants 104, 114, 167, 168, 195–202
Food safety 3, 37, 48, 59, 104, 113, 114, 182, 190, 191, 195, 197
Food sampling methodology 36, 182
Foodstuffs 35–45, 196
Fruits 3, 4, 8, 35, 37, 158, 184, 188, 189

G

Gas chromatography coupled to tandem mass spectrometry (GC-MS/MS) 13–15, 20, 22
Gastric toxicity 114
Gene mutations 123–125, 127, 134, 135
Genotoxicity 121–136

H

Hazards 143, 195, 196
Heterocyclic aromatic amines (HAA) 25–33
Human exposure 12, 158

I

IC_{50} 93, 110
Inductively coupled plasma mass spectrometry (ICP-MS) 35–45, 123, 149, 150, 154, 190
Ingestion models 113–120
Ingestion pathway 158
Inorganic contaminants 35–45
Intake 13, 58, 122, 162, 185, 190, 191, 196, 197
Intestine 168, 169, 171–173, 176
In vitro digestion 122, 123, 125, 158, 161

L

Liquid chromatography 4, 6, 26, 47–56, 72, 82, 174, 190
Liver 12, 86, 90, 91, 167–176

M

Mass spectrometry 6, 13, 26, 47–56, 82, 123, 174, 190, 197
Metabolic activity 103, 104, 108–110, 114

Paula Alvito and Helena Oliveira (eds.), *Food Contaminants Toxicity*, Methods and Protocols in Food Science,
https://doi.org/10.1007/978-1-0716-5096-7,

Micronucleus 124–126, 128, 132–136, 163
Microwave digestion 37, 39, 44, 143, 149
MTT assay 104, 106, 108, 110–111
Mycotoxins 81–99, 104, 113, 123, 167–175, 190

N

Nanomaterials (NMs) 83, 121–136, 141–144, 150, 153, 154
Nanotoxicology 142
New approach methodology (NAM) 122, 125
Nitrosamines 58

O

Oxidative stress...71, 163, 173

P

Persistent organic pollutants (POPs).................... 12, 190
Phenolic compounds extraction 78
Polybrominated diphenyl ethers (PBDEs) 11–22
Potentially toxic elements (PTEs)...................... 141–143, 146–153

Q

Quick, Easy, Cheap, Effective, Rugged, and Safe (QuEChERS) 3–8, 13, 18, 19

R

Reactive oxygen species (ROS) 12, 71, 72, 77, 108, 163
Residues of pesticides.. 6, 8
Root vegetables 57–59

S

Sample analysis 42, 77
Slices.. 167–175
Solid-phase extraction (SPE).............................. 4, 16–18, 20, 22, 26, 27, 48, 49, 53, 55
SPE cleanup...55, 97
Spheroids 114–119

T

TDS food list...183–185, 192
TDS samples............................... 183–185, 187, 189, 190
3D models 113–120
Toxicity 71, 72, 77, 78, 82, 83, 96, 97, 104, 113–120, 122, 141, 143, 148, 158, 168, 169, 173–175, 190
Toxicological potential 158

U

Ultra-high performance liquid chromatography–mass spectrometry 6, 47–56
Urinary biomarkers 196

MIX
Papier aus verantwortungsvollen Quellen
Paper from responsible sources
FSC® C105338

If you have any concerns about our products,
you can contact us on
ProductSafety@springernature.com

In case Publisher is established outside the EU,
the EU authorized representative is:
Springer Nature Customer Service Center GmbH
Europaplatz 3, 69115 Heidelberg, Germany

Printed by Libri Plureos GmbH
in Hamburg, Germany